Samuel Pierpont Langley

The New Astronomy

Samuel Pierpont Langley

The New Astronomy

ISBN/EAN: 9783337396190

Printed in Europe, USA, Canada, Australia, Japan

Cover: Foto ©berggeist007 / pixelio.de

More available books at **www.hansebooks.com**

THE NEW ASTRONOMY

BY

SAMUEL PIERPONT LANGLEY, Ph. D., LL. D.

DIRECTOR OF THE ALLEGHENY OBSERVATORY, MEMBER NATIONAL ACADEMY,
CORRESPONDENT OF THE INSTITUTE OF FRANCE, FELLOW ROYAL
ASTRONOMICAL SOCIETY, ETC., ETC.

Illustrated

BOSTON AND NEW YORK
HOUGHTON, MIFFLIN AND COMPANY
The Riverside Press, Cambridge
1889

PREFACE.

I HAVE written these pages, not for the professional reader, but with the hope of reaching a part of that educated public on whose support he is so often dependent for the means of extending the boundaries of knowledge.

It is not generally understood that among us not only the support of the Government, but with scarcely an exception every new private benefaction, is devoted to "the Old" Astronomy, which is relatively munificently endowed already; while that which I have here called "the New," so fruitful in results of interest and importance, struggles almost unaided.

We are all glad to know that Urania, who was in the beginning but a poor Chaldean shepherdess, has long since become well-to-do, and dwells now in state. It is far less known than it should be that she has a younger sister now among us, bearing every mark of her celestial birth, but all unendowed and portionless. It is for the reader's interest in the latter that this book is a plea.

CONTENTS.

Chapter		Page
I.	Spots on the Sun	1
II.	The Sun's Surroundings	35
III.	The Sun's Energy	70
IV.	The Sun's Energy (*Continued*)	91
V.	The Planets and the Moon	117
VI.	Meteors	175
VII.	Comets	199
VIII.	The Stars	221
	INDEX	253

LIST OF ILLUSTRATIONS.

FIGURE		PAGE
1.	The Sun's Surroundings	4
2.	View of the Sun on Sept. 20, 1870	6
3.	The Sun on Sept. 22, 1870 .	6
4.	The Sun on Sept. 26, 1870	7
5.	The Sun on Sept. 19, 1870 . .	8
6.	The Sun on Sept. 20, 1870	8
7.	The Sun on Sept. 21, 1870	9
8.	The Sun on Sept. 22, 1870	9
9.	The Sun on Sept. 23, 1870	10
10.	The Sun on Sept. 26, 1870	10
11.	Nasmyth's Willow Leaves	11
12.	The Cactus Type	12
13.	Equatorial Telescope and Projection	13
14.	Polarizing Eye-piece	14
15.	Spot of Sept. 21, 1870	15
16.	Spot of March 5, 1873.	15
17.	Sun on March 5, 1873	18
18.	"The Plume" Spot of March 5 and 6, 1873	19
19.	Typical Sun-spot of December, 1873	21
20.	Frost Crystal	23
21.	Cyclone Spot	24
22.	Spot of March 31, 1875	25
23.	Cirrous Cloud	27

LIST OF ILLUSTRATIONS.

FIGURE		PAGE
24.	Spot of March 31, 1875	28
25.	Typical Illustration of Faye's Theory	29
26.	Spot of Oct. 13, 1876	30
27.	Photograph of Edge of Sun	31
28.	Facula	33
29.	Lunar Cone Shadow	36
30.	Track of Lunar Shadow	39
31.	Inner Corona Eclipse of 1869	40
32.	Sketch of Outer Corona, 1869	41
33.	Tacchini's Drawing of Corona of 1870	43
34.	Watson's Naked-eye Drawing of Corona of 1870	44
35.	Photograph showing Commencement of Outer Corona	45
36.	Eclipse of 1857, Drawing by Liais	48
37.	Enlargement of Part of Fig. 38	49
38.	Fac-simile of Photograph of Corona of 1871	51
39.	"Spectres"	54
40.	Outer Corona of 1878	57
41.	Spectroscope Slit and Solar Image	59
42.	Slit and Prominences	59
43.	Tacchini's Chromospheric Clouds	62
44.	Tacchini's Chromospheric Clouds	62
45.	Vogel's Chromospheric Forms	64
46.	Tacchini's Chromospheric Forms	66
47.	Eruptive Prominences	67
48.	Sun-spots and Price of Grain	77
49.	Sun-spot of Nov. 16, 1882, and Earth	80
50.	Greenwich Record of Disturbance of Magnetic Needle, Nov. 16 and 17, 1882	81
51.	Sun-spots and Magnetic Variations	87
52.	Greenwich Magnetic Observations, Aug. 3 and 5, 1872	89
53.	One Cubic Centimetre	93
54.	Pouillet's Pyrheliometer	93
55.	Bernières's great Burning-Glass	103

LIST OF ILLUSTRATIONS. xi

FIGURE	PAGE
56. A "Pour" from the Bessemer Converter	105
57. Photometer-box	108
58. Mouchot's Solar Engine	109
59. Ericsson's new Solar Engine, now in Practical Use in New York	113
60. Saturn	119
61. The Equatorial Telescope at Washington	122
62. Jupiter, Moon, and Shadow	125
63. Three Views of Mars	129
64. Map of Mars	129
65. The Moon	137
66. The Full Moon	141
67. Glass Globe, Cracked	145
68. Plato and the Lunar Alps	149
69. The Lunar Apennines: Archimedes	153
70. Vesuvius and Neighborhood of Naples	157
71. Ptolemy and Arzachel	161
72. Mercator and Campanus	165
73. Withered Hand	168
74. Ideal Lunar Landscape and Earth-shine	169
75. Withered Apple	171
76. Gassendi. Nov. 7, 1867	173
77. The Camp at Mount Whitney	177
78. Vesuvius during an Eruption	183
79. Meteors observed Nov. 13 and 14, 1868, between Midnight and Five o'Clock, A.M.	189
80. Comet of Donati, Sept. 16, 1858	201
81. "A Part of a Comet"	203
82. Comet of Donati, Sept. 24, 1858	205
83. Comet of Donati, Oct. 3, 1858	209
84. Comet of Donati, Oct. 9, 1858	213
85. Comet of Donati, Oct. 5, 1858	217
86. Types of Stellar Spectra	222

LIST OF ILLUSTRATIONS.

FIGURE	PAGE
87. The Milky Way	225
88. Spectra of Stars in Pleiades	231
89. Spectrum of Aldebaran	235
90. Spectrum of Vega	235
91. Great Nebula in Orion	239
92. A Falling Man	243
93. A Flash of Lightning	245

THE NEW ASTRONOMY.

I.

SPOTS ON THE SUN.

THE visitor to Salisbury Plain sees around him a lonely waste, utterly barren except for a few recently planted trees, and otherwise as desolate as it could have been when Hengist and Horsa landed in Britain; for its monotony is still unbroken except by the funeral mounds of ancient chiefs, which dot it to its horizon, and contrast strangely with the crowded life and fertile soil which everywhere surround its borders. In the midst of this loneliness rise the rude, enormous monoliths of Stonehenge, — circles of gray stones, which seem as old as time, and were there, as we now are told, the temple of a people which had already passed away, and whose worship was forgotten, when our Saxon forefathers first saw the place.

In the centre of the inner circle is a stone which is believed once to have been the altar; while beyond the outmost ring, quite away to the northeast upon the open plain, still stands a solitary stone, set up there evidently with some special object by the same unknown builders. Seen under ordinary circumstances, it is difficult to divine its connection with the others; but we are told that once in each year, upon the morning of the longest day, the level shadow of this distant, isolated stone is projected at sunrise to the very centre of the ancient sanctuary, and falls just upon the altar. The primitive man who devised this was both astronomer and priest, for he not only adored the risen god whose first beams brought him light and warmth, but he could mark its place, and though utterly ignorant of its nature, had evidently learned enough of its motions to embody his simple astronomical knowledge in a record so exact and so enduring that though his very memory has gone, common men are still interested in it; for, as I learned when viewing the scene, people are accustomed to come from all the surrounding country, and pass in this desolate spot the short night preceding the longest day of the year, to see the shadow touch the altar at the moment of sunrise.

Most great national observatories, like Greenwich or Washington, are the perfected development of that kind of astronomy of which the builders of Stonehenge represent the infancy. Those primitive men could know where the sun would rise on a certain day, and make their observation of its place, as we see, very well, without knowing anything of its physical nature. At Greenwich the moon has been observed with scarcely an intermission for one hundred and fifty years, but we should mistake greatly did we suppose that it was for the purpose of seeing what it was made of, or of making discoveries in it. This immense mass of Greenwich observations is for quite another purpose, — for the very practical purpose of forming the lunar tables, which, by means of the moon's place among the stars, will tell the navigator in distant oceans where he is, and conduct the fleets of England safely home.

In the observatory at Washington one may see a wonderfully exact instrument, in which circles of brass have replaced circles of stone, all so bolted between massive piers that the sun can be observed by it but once daily, as it crosses the meridian. This instrument is the completed attainment along that long line of progress in one direction, of which the solitary stone at Stonehenge marks the initial step, — the attainment, that is, purely of precision of measurement; for the astronomer of to-day can still use his circles for the special purpose of fixing the sun's place in the heavens, without any more knowledge of that body's chemical constitution than had the man who built Stonehenge.

Yet the object of both is, in fact, the same. It is true that the functions of astronomer and priest have become divided in the advance of our modern civilization, which has committed the special cultivation of the religious aspect of these problems to a distinct profession; while the modern observer has possibly exchanged the emotions of awe and wonder for a more exact knowledge of the equinox than was possessed by his primitive brother, who both observed and adored. Still, both aim at the common end, not of learning what the sun is made of, but of where it will be at a certain moment; for the prime object of astronomy, until very lately indeed, has still been to say *where* any heavenly body is, and not *what* it is. It is this precision of measurement, then, which has always — and justly — been a paramount object of this oldest of the sciences, not only as a good in itself, but as leading to great ends; and it is this which the poet of Urania has chosen rightly to note as its characteristic, when he says, —

> "That little Vernier, on whose slender lines
> The midnight taper trembles as it shines,
> Tells through the mist where dazzled Mercury burns,
> And marks the point where Uranus returns."

But within a comparatively few years a new branch of astronomy has arisen, which studies sun, moon, and stars for what they are in themselves, and in relation to ourselves. Its study of the sun, beginning with its external features (and full of novelty and interest, even, as regards those), led to the further inquiry as to what it was made of, and then to finding the unexpected relations which it bore to the earth and our own daily lives on it, the conclusion being that, in a physical sense, it made us and re-creates us, as it were, daily, and that the knowledge of the intimate ties which unite man with it brings results of the most practical and important kind, which a generation ago were unguessed at.

This new branch of inquiry is sometimes called Celestial Physics, sometimes Solar Physics, and is sometimes more rarely referred to as the New Astronomy. I will call it here by this title, and try to tell the reader something about it which may interest him, beginning with the sun.

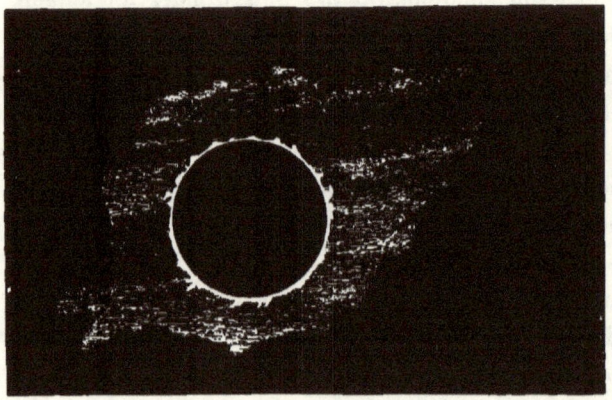

FIG. 1. — THE SUN'S SURROUNDINGS.

The whole of what we have to say about the sun and stars presupposes a knowledge of their size and distance, and we may take it for granted that the reader has at some time or another heard such statements as that the moon's distance is two hundred and forty thousand miles, and the sun's ninety-three million (and very probably has forgotten them again as of no practical concern). He will not be offered here the kind of statistics which he would expect in a college text-book; but we must linger a moment on the threshold of our subject — the nature of these bodies — to insist on the real meaning of such figures as those just quoted. We are accustomed to look on the sun and moon as far off together in the sky; and though we know the sun is greater, we are apt to think of them vaguely as things of a common order of largeness, away among the stars. It would be safe to say that though nine out of ten readers have learned that the sun is larger than the moon, and, in fact, larger than the earth itself, most of them do not at all realize that the difference is so enormous that if we could hollow out the sun's globe and place the earth in the centre, there would still be so much room that the moon might go on moving in her present orbit at two hundred and forty thousand miles from the earth, — *all within the globe of the sun itself,* — and have plenty of room to spare.

As to the distance of ninety-three million miles, a cannon-ball would travel it in about fifteen years. It may help us to remember that at the speed attained by the Limited Express on our railroads a train which had left the sun for the earth when the "Mayflower" sailed from Delftshaven with the Pilgrim Fathers, and which ran at that rate day and night, would in 1887 still be a journey of some years away from its terrestrial station. The fare at the customary rates, it may be remarked, would be rather over two million five hundred thousand dollars, so that it is clear that we should need both money and leisure for the journey.

Perhaps the most striking illustration of the sun's distance is given by expressing it in terms of what the physiologists would call velocity of nerve transmission. It has been found that sensation is not absolutely instantaneous, but that it occupies a very minute time in travelling along the nerves; so that if a child puts its finger into the candle, there is a certain almost inconceivably small space of time, say the one-hundredth of a second, before he feels the heat. In case, then, a child's arm were long enough to touch the sun, it can be calculated from this known rate of transmission that the infant would have to live to be a man of over a hundred before it knew that its fingers were burned.

Trying with the help of these still inadequate images, we may get some idea of the real size and distance of the sun. I could wish not to have to dwell upon such figures, that seem, however, indispensable; but we are now done with these, and

FIG. 2.—VIEW OF THE SUN ON SEPT. 20, 1870. FIG. 3.—THE SUN ON SEPT. 22, 1870.
(FROM A PHOTOGRAPH.)

are ready to turn to the telescope and see what the sun itself looks like.

The sun, as we shall learn later, is a star, and not a particularly large star. It is, as has been said, "only a private in the host of heaven," but it is one of that host; it is one of those glittering points to which we have been brought near. Let us keep in mind, then, from the first, what we shall see confirmed later, that there is an essentially similar constitution in them all, and not forget that when we study the sun, as we now begin to do, we are studying the stars also.

If we were called on to give a description of the earth and all that is on it, it would be easily understood that the task was impossibly great, and that even an account of its most striking general features might fill volumes. So it is with the sun; and we shall find that in the description of the general character of its immediate surface alone, there is a great deal to be told. First, let us look at a little conventional representation (Fig. 1), as at a kind of outline of the unknown regions we are about to explore. The circle represents the Photosphere, which is simply what the word implies, that "sphere" of "light" which we have daily before our eyes, or which we can study with the telescope. Outside this there is a thin envelope, which rises here and there into irregular prominences, some orange-scarlet, some rose-pink. This is the Chromosphere, a thin shell, mainly of crimson and scarlet tints, invisible even to the telescope except at the time of a total eclipse, when alone its true colors are discernible, but seen as to its form at all times by the spectroscope. It is always there, not hidden in any way, and yet not seen, only because it is overpowered by the intenser brilliancy of the Photosphere, as a glowworm's shine would be if it were put beside an electric light.

FIG. 4. — THE SUN ON SEPT. 26, 1870.

Outside all is the strange shape, which represents the mysterious Corona, seen by the naked eye in a total eclipse, but at all other times invisible even to telescope and spectroscope, and of whose true nature we are nearly ignorant from lack of opportunity to study it.

Disregarding other details, let us carry in our minds the three main divisions, — the Photosphere, or daily visible surface of the sun, which contains nearly all its mass or substance; the Chromosphere; and the unsubstantial Corona, which is nevertheless larger than all the rest. We begin our examination with the Photosphere.

There are records of spots having been seen with the naked eye before the invention of the telescope, but they were supposed to be planets passing between us and the surface; and the idea that the sun was pure fire, necessarily immaculate, was taught by the professors of the Aristotelian philosophy in mediæval schools, and regarded almost as an article of religious faith. We can hardly conceive, now, the shock of the first announcement that spots were to be found on the sun, but the notion partook in contemporary minds at once of the absurd and the

FIG. 5.— SEPT. 19, 1870.   FIG. 6.— SEPT. 20, 1870.
(ENGRAVED FROM A PHOTOGRAPH BY RUTHERFURD.)

impious; and we notice here, what we shall have occasion to notice again, that these physical discoveries from the first affect men's thoughts in unexpected ways, and modify their scheme of the moral universe as well as of the physical one.

Very little indeed was added to the early observations of Fabricius and Galileo until a time within the remembrance of many of us; for it is since the advent of the generation now on the stage that nine-tenths of the knowledge of the subject has been reached.

Let us first take a general view of the sun, and afterward study it in detail. What we see with a good telescope in this general view is something like this. Opposite are three successive views (Figs. 2, 3, 4) taken on three successive days,—quite authentic portraits, since the sun himself made them; they being, in fact, projected telescopic images which have been fixed for us by photography, and then exactly reproduced by the engraver. The first was taken (by Mr. Rutherfurd, of New York) on the

FIG. 7.—SEPT. 21, 1870.

FIG. 8.—SEPT. 22, 1870.

20th of September, 1870, when a remarkably large spot had come into view. It is seen here not far from the eastern edge (the left hand in the engraving), and numerous other spots are also visible. The reader should notice the position of these, and then on turning to the next view (Fig. 3, taken on September 22d) he will see that they have all shifted their places, by a common motion toward the west. The great spot on the left has now got well into view, and we can see its separate parts; the group which was on the left of the centre has got a little to the right of it, and so on. From the common motion of them all, we might suspect that the sun was turning round on an axis like the earth, carrying the spots with it; and as we continue to observe, this suspicion becomes certainty. In the third view (Fig. 4), taken on September 26th, the spot we first saw on the left has travelled more than half across the disk, while others we saw on September 20th have approached to the right-hand edge or passed wholly out of sight behind it. The sun does rotate, then, but in twenty-five or twenty-six of our days,—I say twenty-five *or* twenty-six, because (what is very extraordinary) it does not turn all-of-a-piece like the earth, but some parts revolve faster than others,—not only faster in feet and inches, but in the number of turns,—just as though the rim of a carriage wheel were to make more revolutions in a mile than the spokes, and the spokes more than the hub. Of course no solid wheel could so turn without wrenching itself in pieces, but that the great solar wheel does, is incontestable; and this alone is a convincing proof that the sun's surface is not solid, but liquid or gaseous.

But let us return to the great spot which we saw coming round the eastern edge. Possibly the word "great" may seem misapplied to what was but the size of a pin-head in the first engraving, but we must remember that the disk of the sun there shown is in reality over 800,000 miles in diameter. We

 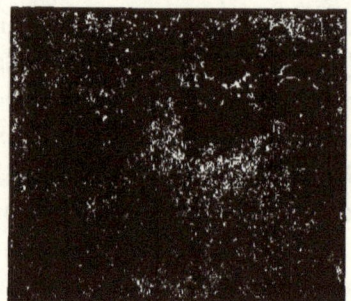

FIG. 9.— SEPT. 23, 1870. FIG. 10.— SEPT. 26, 1870.

shall soon see whether this spot deserves to be called "great" or not.

Next we have six enlarged views of it on the 19th, 20th, 21st, 22d, 23d, and 26th. On the 19th it is seen very near the eastern limb, showing like a great hole in the sun, and foreshortened as it comes into view around the dark edge; for the edge of the sun is really darker than the central parts, as it is shown here, or as one may see even through a smoked glass by careful attention. On the 20th we have the edge still visible, but on the 21st the spot has advanced so far that the edge cannot be shown for want of room. We see distinctly the division of the spot into the outer shades which constitute the penumbra, and

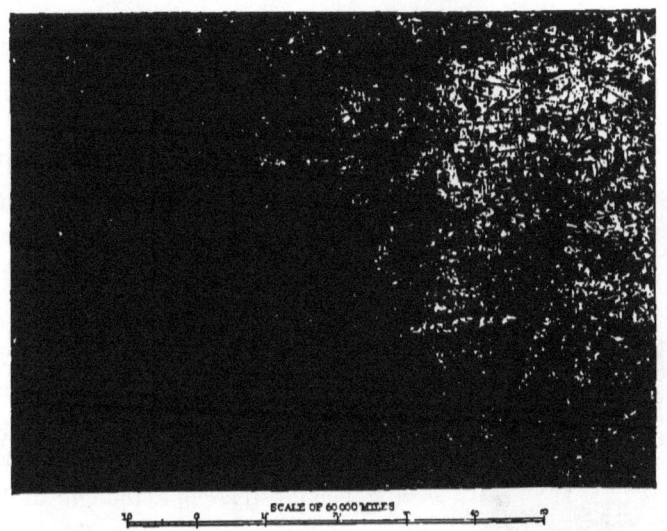

FIG. 11.—NASMYTH'S WILLOW LEAVES. (FROM HERSCHEL'S "OUTLINES OF ASTRONOMY.")

the inner darker ones which form the umbra and nucleus. We notice particularly in this enlarged view, by comparing the appearances on the 21st, 22d, and 23d, that the spot not only turns with the sun (as we have already learned), but moves and changes within itself in the most surprising way, like a terrestrial cloud, which not only revolves with the rest of the globe, but varies its shape from hour to hour. This is seen still more plainly when we compare the appearance on the 23d with that on the 26th, only three days later, where the process has begun by which the spot finally breaks up and forever disappears. On looking at all this, the tremendous scale on which the action occurs must be borne in mind. On the 21st, for instance, the umbra, or dark central hole, alone was large enough to let the whole globe of our own earth drop in without touching the sides! We shall have occasion to recur to this view of the 21st September again.

In looking at this spot and its striking changes, the reader must not omit to notice, also, a much less obvious feature, — the vaguely seen mottlings which show all over the sun's surface, both quite away from the spots and also close to them, and which seem to merge into them.

FIG. 12. — THE CACTUS TYPE.
(FROM SECCHI'S "LE SOLEIL.")

I think if we assign one year rather than another for the birth of the youthful science of solar physics, it should be 1861, when Kirchhoff and Bunsen published their memorable research on Spectrum Analysis, and when Nasmyth observed what he called the "willow-leaf" structure of the solar surface (see Fig. 11). Mr. Nasmyth, with a very powerful reflecting telescope, thought he had succeeded in finding what these faint mottlings really are composed of, and believed that he had discovered in them some most extraordinary things. This is what he thought he saw: The whole sun is, according to him, covered with huge bodies of most definite shape, that of the oblong willow leaf, and of enormous but uniform size; and the faint mottlings the reader has just noticed are, according to him, made up of these. "These," he says, "cover the whole disk of the sun (except in the space occupied by the spots) in countless millions, and lie crossing each other in every imaginable direction." Sir John Herschel took a particular interest in the supposed discovery, and, treating it as a matter of established fact, proceeded to make one of the most amazing suggestions in explanation that ever came from a scientific man of deserved eminence. We must remember how much there is unknown in the sun still, and what a great mystery even yet overhangs many of our relations to that body which maintains our own vital action, when we read the following words, which are Herschel's own. Speaking of these supposed spindle-shaped monsters, he says:

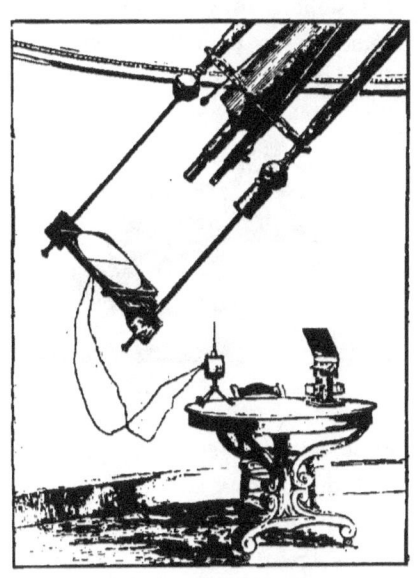

FIG. 13.—EQUATORIAL TELESCOPE AND PROJECTION.

"The exceedingly definite shape of these objects, their exact similarity to one another, and the way in which they lie across and athwart each other, — all these characters seem quite repugnant to the notion of their being of a vaporous, a cloudy, or a fluid nature. Nothing remains but to consider them as separate and independent sheets, flakes, or scales, having some sort of solidity. And these . . . are evidently *the immediate sources of the solar light and heat*, by whatever mechanism or whatever processes they may be enabled to develop, and as it were elaborate, these elements from the bosom of the non-luminous fluid in which they appear to float. Looked at in this point of view, we cannot refuse to regard them as *organisms* of some peculiar and amazing kind; and though it would be too daring to speak of such organization as partaking of the nature of life, yet we do know that vital action is competent to develop at once heat and light and electricity."

Such are his words; and when we consider that each of these solar inhabitants was supposed to extend about two hundred by one thousand miles upon the surface of the fiery ocean, we may subscribe to Mr. Proctor's comment, that "Milton's picture of him who on the fires of hell 'lay floating many a rood,' seems tame and commonplace compared with Herschel's conception of these floating monsters, the least covering a greater space than the British Islands."

I hope I may not appear wanting in respect for Sir John Herschel — a man whose memory I reverence — in thus citing views which, if his honored life could have been prolonged, he would have abandoned. I do so because nothing else can so forcibly illustrate the field for wonder and wild conjecture solar physics presented even a few years ago; and its supposed connection with that "Vital Force," which was till so lately accepted by physiology, serves as a kind of landmark on the way we have come.

FIG. 14. — POLARIZING EYE-PIECE.

This new science of ours, then, youthful as it is, has already had its age of fable.

After a time Nasmyth's observation was attributed to imperfect definition, but was not fairly disproved. He had, indeed, a basis of fact for his statement, and to him belongs the credit of first pointing out the existence of this minute structure, though he mistook its true character. It will be seen later how the real forms might be mistaken for leaves, and *in certain particular cases* they certainly do take on a very leaf-like appearance. Here is a drawing (Fig. 12) which Father Secchi gives of some of them in the spot of April 14, 1867, and which he compares to a branch of cactus. He remarks somewhere else that they resemble a crystallization of sal-ammoniac, and calls them veils

FIG. 15.— SPOT OF SEPT. 21, 1870. (REDUCED FROM AN ORIGINAL DRAWING BY S. P. LANGLEY.)

FIG. 16.— SPOT OF MARCH 5, 1873. (REDUCED FROM AN ORIGINAL DRAWING BY S. P. LANGLEY.)

of most intricate structure. This was the state of our knowledge in 1870, and it may seem surprising that such wonderful statements had not been proved or disproved, when they referred to mere matters of observation. But direct observation is here very difficult on account of the incessant tremor and vibration of our own atmosphere.

The surface of the sun may be compared to an elaborate engraving, filled with the closest and most delicate lines and hatchings, but an engraving which during ninety-nine hundredths of the time can only be seen across such a quivering mass of heated air as makes everything confused and liable to be mistaken, causing what is definite to look like a vaguely seen mottling. It is literally true that the more delicate features we are about to show, are only distinctly visible even by the best telescope during less than one-hundredth of the time, coming out as they do in brief instants when our dancing air is momentarily still, so that one who has sat at a powerful telescope all day is exceptionally lucky if he has secured enough glimpses of the true structure to aggregate five minutes of clear seeing, while at all other times the attempt to magnify only produces a blurring of the image. This study, then, demands not only fine telescopes and special optical aids, but endless patience.

My attention was first particularly directed to the subject in 1870 (shortly after the regular study of the Photosphere was begun at the Allegheny Observatory by means of its equatorial telescope of thirteen inches' aperture), with the view of finding out what this vaguely seen structure really is. Nearly three years of constant watching were given to obtain the results which follow. The method I have used for it is indicated in the drawing (Fig. 13), which shows the preliminary step of projecting the image of the sun directly upon a sheet of paper, divided into squares and attached to the eye-end of a great equatorial telescope. When this is directed to the sun in a darkened dome, the solar picture is formed upon the paper as in a camera obscura, and this picture can be made as large or as small as we please by varying the lenses which project it. As the sun moves along in the sky, its image moves across the paper; and as we can observe how long the whole sun (whose diameter in miles is known) takes to cross, we can find how many miles correspond to the time it is in crossing one of the squares, and so get the scale of the future drawing, and the true size in miles of the spot we are about to study. Then a piece of clock-work attached to the telescope is put in motion, and it begins to follow the sun in the sky, and the spot appears fixed on the paper. A tracing of the spot's outline is next made, but the finer details are not to be observed by this method, which is purely preliminary, and only for the purpose of fixing the scale and the points of the compass (so to speak) on the sun's face. The projecting apparatus is next removed and replaced by the polarizing eye-piece. Sir William Herschel used to avoid the blinding effects of the concentrated solar light by passing the rays through ink and water, but the phenomena of "polarization" have been used to better advantage in modern apparatus. This instrument, one of the first of its kind ever constructed, and in which the light is polarized with three successive reflections through the three tubes seen in the drawing (Fig. 14), was made in Pittsburgh as a part of the gift of appa-

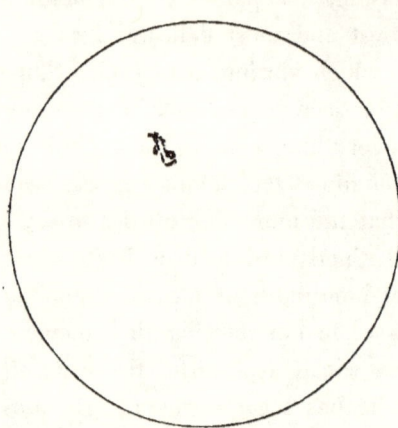

FIG. 17.—SUN ON MARCH 5, 1873. (FROM A DRAWING BY S. P. LANGLEY.)

ratus by one of its citizens to the Observatory, and has been most useful. By its aid the eye can be safely placed where the concentrated heat would otherwise melt iron. In practice I have often gazed through it at the sun's face without intermission from four to five hours, with no more fatigue or harm to the eye than in reading a book. By its aid the observer fills in the outline already projected on the paper.

The photograph has transported us already so near the sun's surface that we have seen details there invisible to the naked eye. We have seen that what we have called "spots" are indeed regions whose actual vastness surpasses the vague immensity of a dream, and it will not cause surprise that in them is a temperature which also surpasses greatly that of the hottest furnace. We shall see later, in fact, that the whole surface is composed largely of metals turned into vapor in this heat, and that if we could indeed drop our great globe itself upon the sun, it would be dissipated as a snow-flake.

FIG. 18. — "THE PLUME" SPOT OF MARCH 5 AND 6, 1873. (FROM AN ORIGINAL DRAWING BY S. P. LANGLEY.)

Now, we cannot suppose this great space is fully described when we have divided it into the penumbra, umbra, and nucleus, or that the little photograph has shown us all there is, and we rather anticipate that these great spaces must be filled with curious things, if we could get near enough to see them. We cannot advantageously enlarge our photograph further; but if we could really come closer, we should have the nearer view that the work at Allegheny, I have just alluded to, now affords. The drawing (Fig. 15) of the central part of the same great spot, already cited, was made on the 21st of September, 1870, and may be compared with the photograph of that day. We have now a greatly more magnified view than before, but it is not blurred by the magnifying, and is full of detail. We have been brought within two hundred thousand miles of the sun, or rather less than the actual distance of the moon, and are seeing for ourselves what was a few years since thought out of the reach of any observer. See how full of intricate forms that void, black, umbral space in the photograph has become! The penumbra is filled with detail of the strangest kind, and there are two great "bridges," as they are called, which are almost wholly invisible in the photograph. Notice the line in one of the bridges which follows its sinuosities through its whole length of twelve thousand miles, making us suspect that it is made up of smaller parts as a rope is made up of cords (as, in fact, it is); and look at the end, where the cords themselves are unravelled into threads fine as threads of silk, and these again resolved into finer fibres, till in more and more web-like fineness it passes beyond the reach of sight! I am speaking, however, here rather of the wonderful original, as I so well remember it, than of what my sketch or even the engraver's skill can render.

Next we have quite another "spot" belonging to another year (1873). First, there is a view (Fig. 17) of the sun's disk with the spot on it (as it would appear in a small telescope), to show its relative size, and then a larger drawing of the spot itself (Fig. 16), on a scale of twelve thousand miles to the inch, so that the region shown to the reader's eyes, though but a "spot" on the sun, covers an area of over one billion square miles, or more than five times the entire surface of the earth, land, and water. To help us to conceive its vastness, I have drawn in one corner the continents of North and South America

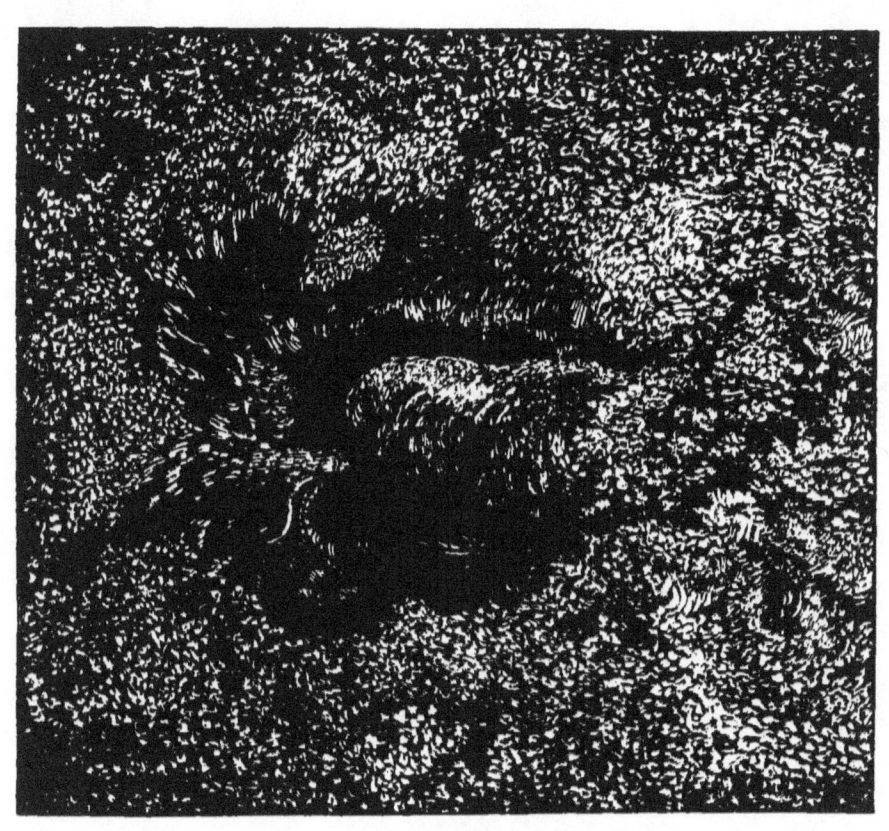

FIG. 19. — TYPICAL SUN SPOT OF DECEMBER, 1873.
(REDUCED FROM AN ORIGINAL DRAWING BY S. P. LANGLEY.)

on the same scale as the "spot." Notice the evidence of solar whirlwinds and the extraordinary "plume" (Fig. 16), which is a something we have no terrestrial simile for. The appearance of the original would have been described most correctly by such incongruous images as "leaf-like" and "crystalline" and "flame-like;" and even in this inadequate sketch there may remain some faint suggestion of the appearance of its wonderful arche-

FIG. 20. — FROST CRYSTAL.

type, which was indeed that of a great flame leaping into spires and viewed through a window covered with frost crystals. Neither "frost" nor "flame" is really there, but we cannot avoid this seemingly unnatural union of images, which was fully justified by the marvellous thing itself. The reader must bear in mind that the whole of this was actually in motion, not merely turning with the sun's rotation, but whirling and shifting within itself, and that the motion was in parts occasionally probably as high as fifty miles per second, — per *second*, remember, not per hour, — so that it changed under the gazer's eyes. The hook-shaped prominence in the lower part (actually larger than the United States) broke up and disappeared in about twenty minutes, or while the writer was engaged in drawing it. The imagination is confounded in an attempt to realize to itself the true character of such a phenomenon.

On page 19 is a separate view of the plume (Fig. 18), a fac-simile of the original sketch, which was made with the eye at the telescope. The pointed or flame-like tips are not a very common form, the terminals being more commonly clubbed, like those in Father Secchi's "branch of cactus" type given on page 12. It must be borne in mind, too, if the drawing does not seem to contain all that the text implies, that there were but a few minutes in which to attempt to draw, where even a skilled draughtsman might have spent hours on the details momentarily visible, and that much must be left to memory. The writer's note-book at the time contains an expression of despair at his utter inability to render most of what he saw.

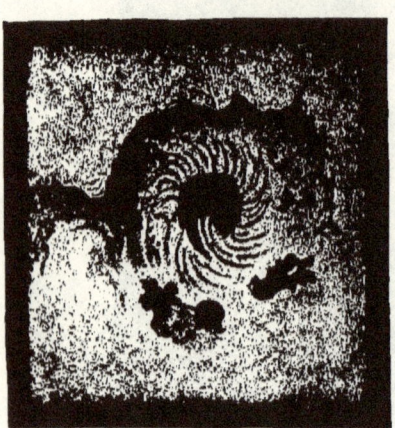

FIG. 21. — CYCLONE SPOT. (DRAWN BY FATHER SECCHI.)

Let us now look at another and even more wonderful example. Fig. 19 shows part of a great spot which the writer drew in December, 1873, when the rare coincidence happened of a fine spot and fine terrestrial weather to observe it in. In this, as well as in the preceding drawing, the pores which cover the sun's surface by millions may be noted. The luminous dots which divide them are what Nasmyth imperfectly saw, but we are hardly more able than he to say what they really are. Each of these countless "dots" is larger than England, Scotland, and Ireland together! The wonderful "crystalline" structure in the centre cannot be a real crystal, for it is ten times the area of Europe, and changed slowly while I drew it; but the reader may be sure that its resemblance to some crystallizations has not been in the least exaggerated. I have sought to study various actual crystals for comparison, but found none quite satisfactory. That of sal-ammoniac in some remote way resembles it, as Secchi says; but perhaps the frost crystals on a window-pane are better. Fig. 20 shows one selected among several windows I had photographed in a preceding winter, which has some suggestions of the so-called crystalline spot-forms in it, but which lacks the filamentary thread-like components presently described.

FIG. 22.—SPOT OF MARCH 31, 1873. (FROM AN ORIGINAL DRAWING BY S. P. LANGLEY.)

Of course the reader will understand that it is given as a suggestion of the appearance merely, and that no similarity of nature is meant to be indicated.

There were wonderful fern-like forms in this spot, too, and an appearance like that of pine-boughs covered with snow; for, strangely enough, the intense whiteness of the solar surface in the best telescopes constantly suggests cold. I have had the same impression vividly in looking at the immense masses of molten-white iron in a great puddling-furnace. The salient feature here is one very difficult to see, even in good telescopes, but one which is of great interest. It has been shown in the previous drawings, but we have not enlarged on it. Everywhere in the spot are long white threads, or filaments, lying upon one another, tending in a general sense toward the centre, and each of which grows brighter toward its inner extremity. These make up, in fact, as we now see, the penumbra, or outer shade, and the so-called "crystal" is really affiliated to them. Besides this, on closer looking we see that the inner shade, or umbra, and the very deepest shades, or nuclei, are really made of them too. We can look into the dark centre, as into a funnel, to the depth of probably over five thousand miles; but as far as we may go down we come to no liquid or solid floor, and see only volumes of whirling vapor, disposed not vaguely like our clouds, but in the singularly definite, fern-like, flower-like forms which are themselves made of these "filaments," each of which is from three to five thousand miles long, and from fifty to two hundred miles thick, and each of which (as we saw in the first spot) appears to be made up like a rope of still finer and finer strands, looking in the rare instants when irradiation makes an isolated one visible, like a thread of gossamer or the finest of cobweb. These suggest the fine threads of spun glass; and here there is something more than a mere resemblance of form, for both appear to have one causal feature in common, due to a viscous or "sticky" fluid; for there is much reason to believe that the solar atmosphere, even where thinner than our own air, is rendered viscous by the enormous heat, and owes to this its tendency to pull out in strings in common with such otherwise dissimilar things as honey, or melted sugar, or melted glass.

We may compare these mysterious things, the filaments, to long grasses growing in the bed of a stream, which show us the direction and the eddies of the current. The likeness holds in more ways than one. They are not lying, as it were, flat upon the surface of the water, but *within* the medium; and they do not stretch along in any one plane, but they bend down and up. Moreover, they are, as we see, apparently rooted at one end, and their tips rise above the turbid fluid and grow brighter as they are lifted out of it. But perhaps the most significant use of the comparison is made if we ask whether the stream is moving in an eddy like a whirlpool or boiling up from the ground. The question in other words is, "Are these spots themselves the sign

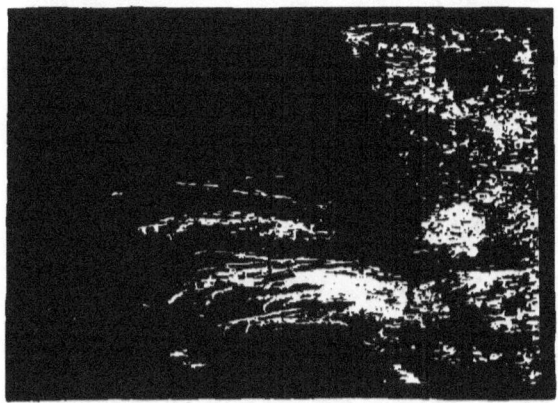

FIG. 23.—CIRROUS CLOUD. (FROM A PHOTOGRAPH.)

of a mere chaotic disturbance, or do they show us by the disposition of these filaments that each is a great solar maelstrom, carrying the surface matter of the sun down into its body? or, finally, are they just the opposite, — something comparable to fiery fountains or volcanoes on the earth, throwing up to the surface the contents of the unknown solar interior?"

Before we try to answer this question, let us remember that the astonishing rapidity with which these forms change, and still more the fact that they do not by any means always change by a bodily removal of one part from another, but by a dissolving away and a fading out into invisibility, like the melting of a cloud into thin air, — let us remember that all this assimilates them to something cloud-like and vaporous, rather than crystalline, and that, as we have here seen, we can ourselves pronounce from such results of recent observation that these are not lumps of scoriæ floating on the solar furnace (as some have thought them), and still less, literal crystals. We can see for ourselves, I believe, that so far there is no evidence here of any solid, or even liquid, but that the surface of the sun is purely vaporous. Fig. 23 shows a cirrous cloud in our own atmosphere, caught for us by photography, and which the reader will find it interesting to compare with the apparently analogous solar cloud-forms.

FIG. 24. — SPOT OF MARCH 31, 1873. (FROM AN ORIGINAL DRAWING BY S. P. LANGLEY.)

"Vaporous," we call them, for want of a better word, but without meaning that it is like the vapor of our clouds. There is no exact terrestrial analogy for these extraordinary forms, which are in fact, as we shall see later, composed of iron and other metals — not of solid iron nor even of liquid, but iron heated beyond even the liquid state to that of iron-steam or vapor.

With all this in mind, let us return to the question, "Are the spots, these gigantic areas of disturbance, comparable to whirlpools or to volcanoes?" It may seem unphilosophical to assume that they are one or the other, and in fact they may possibly be neither; but it is certain that the surface of the sun would soon cool from its enormous temperature, if it were not supplied with fresh heat, and it is almost certain that this heat is drawn from the interior. As M. Faye has pointed out,[1] there must be a circulation up and down, the cooled products being carried within, heated and brought out again, or the sun would, however hot, grow cold outside; and, what is of interest to us, the earth would grow cold also, and we should all die. No one, I believe, who has studied the subject, will contradict the statement that if the sun's surface were absolutely cut off from any heat supply from the interior, organic life in general upon the earth (and our own life in particular) would cease much within a month.

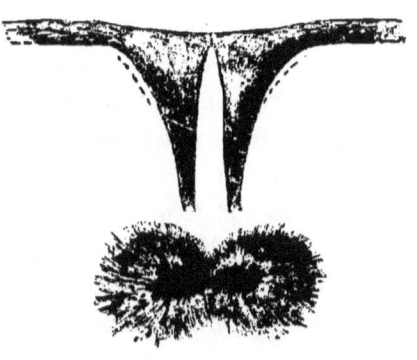

FIG. 25. — TYPICAL ILLUSTRATION OF FAYE'S THEORY.

This solar circulation, then, is of nearly as much consequence to us as that of our own bodies, if we but knew it; and now let us look at the spots again with this in mind.

Fig. 21 shows a drawing by Father Secchi of a spot in 1854; and it is, if unexaggerated, quite the most remarkable case of distinct cyclonic action recorded. I say "if unexaggerated" because there is a strong tendency in most designers to select what is striking in a spot, and to emphasize that unduly, even when there is no conscious disposition to alter. Every one who sketches may see a similar unconscious tendency in himself or herself, shown in a disposition to draw all the mountains and hills too high, — a tendency on which Ruskin, I think, has remarked. In drawings of the sun there is a strong temptation to exaggerate these circular forms, and we must not forget this in making up the evidence. There is great need of caution, then, in receiving such representations; but there certainly are forms which seem to be clearly due to cyclonic action. They are

---

[1] To Mr. Herbert Spencer must be assigned the earliest suggestion of the necessity of such a circulation.

usually scattered, however, through larger spots, and I have never, in all my study of the sun, seen one such complete type of the cyclone spot as that first given from Secchi. Instances where spots break up into numerous subdivisions by a process of "segmentation" under the apparent action of separate whirlwinds are much more common. I have noticed, as an apparent effect of this segmentation, what I may call the "honeycomb structure" from its appearance with low powers, but which with higher ones turns out to be made up of filamentary masses disposed in circular and ovoid curves, often apparently overlying one another, and frequently presenting a most curious resemblance to vegetable forms, though we appear to see the real agency of whirlwinds in making them. I add some transcripts of my original pencil memoranda themselves, made with the eye at the telescope, which, though not at all finished drawings, may be trusted the more as being quite literal transcripts at first hand.

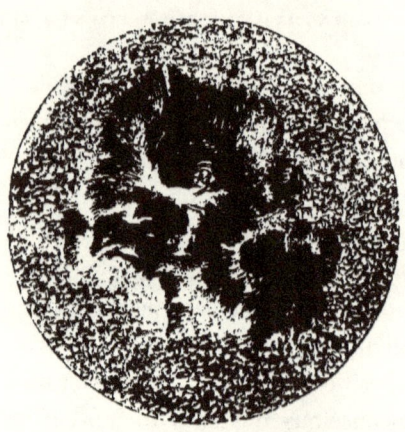

FIG. 26. — SPOT OF OCT. 13, 1876. (FROM ORIGINAL DRAWING BY S. P. LANGLEY.)

Figs. 22 and 24, for instance, are two sketches of a little spot, showing what, with low powers, gives the appearance I have called the honeycomb structure, but which we see here to be due to whirls which have disposed the filaments in these remarkable forms. The first was drawn at eleven in the forenoon of March 31, 1875, the second at three in the afternoon of the same day. The scale of the drawing is fifteen thousand miles to the inch, and the changes in this little spot in these few hours imply a cataclysm compared with which the disappearance of the American continent from the earth's surface would be a trifle.

The very act of the solar whirlwind's motion seemed to pass before my eyes in some of these sketches; for while drawing them as rapidly as possible, a new hole would be formed where there was none before, as if by a gigantic invisible auger boring downward.

M. Faye, the distinguished French astronomer, believes that, owing to the fact that different zones of the sun rotate faster than others, whirlwinds analogous to our terrestrial cyclones, but on a vaster scale, are set in motion, and suck down the cooled vapors of the solar surface into its interior, to be heated and returned again, thus establishing a circulation which keeps the surface from cooling down. He points out that we should not conclude that these whirlwinds are not acting everywhere, merely because our bird's-eye view does not always show them. We see that the spinning action of a whirlpool in water becomes more marked as we go below the surface, which is comparatively undisturbed, and we often see one whirl break up into several minor ones, but all sucking downward and never upward. According to M. Faye, something very like this takes place on the sun, and in Fig. 25 he gives this section to show what he believes to occur in the case of a spot which has "segmented," or divided into two, like

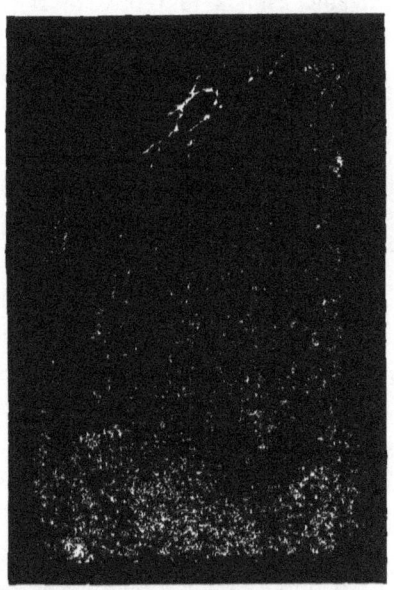

FIG. 27.—PHOTOGRAPH OF EDGE OF SUN. (BY PERMISSION OF WARREN DE LA RUE, LONDON.)

the one whose (imaginary) section is shown above it. This theory is to be considered in connection with such drawings as we have just shown, which are themselves, however, no way dependent on theory, but transcripts from Nature.

I do not here either espouse or oppose the "cyclonic" theory, but it is hardly possible for any one who has been an eyewitness of such things to refuse to regard some such disturbance as a real and efficient cause in such instances as this.

Fig. 26, on nearly the same scale as the last, shows a spot which was seen on Oct. 13, 1876. It looked at first, in the telescope, like two spots without any connection; then, as vision improved and higher powers were employed, the two were seen to have a subtle bond of union, and each to be filled with the most curious foliage-forms, which I could only indicate in the few moments that the good definition lasted. The reader may be sure, I think, that there is no exaggeration of the curious shapes of the original; for I have been so anxious to avoid the overstatement of curvature that the error is more likely to be in the opposite direction.

We must conclude that the question as to the cyclonic hypothesis cannot yet be decided, though the probabilities from telescopic evidence at present seem to me on the whole in favor of M. Faye's remarkable theory, which has the great additional attraction to the student that it unites and explains numerous other quite disconnected facts.

Turning now to the other solar features, let us once more consider the sun as a whole. Fig. 27 is a photograph taken from a part of the sun near its edge. We notice on it, what we see on every careful delineation of the sun, that its general surface is not uniformly bright, but that it grows darker as we approach the edge, where it is marked by whiter mottlings called faculæ, "something in the sun brighter than the sun itself," and looking in the enlarged view which we present of one of them

(Fig. 28), as if the surface of partly cooled metal in a caldron had been broken into fissures showing the brighter glow beneath. These "faculæ," however, are really above the solar surface, not below it, and what we wish to direct particular attention to is that darkening toward the edge which makes them visible.

This is very significant, but its full meaning may not at first be clear. It is owing to an atmosphere which surrounds the sun, as the air does the earth. When we look horizontally through our own air, as at sunrise and sunset, we gaze through greater thicknesses of it than when we turn our eyes to the zenith. So when we look at the edge of the sun, the line of sight passes through greater depths of this solar atmosphere, and it dims the light shining behind it more than at the centre, where it is thin.

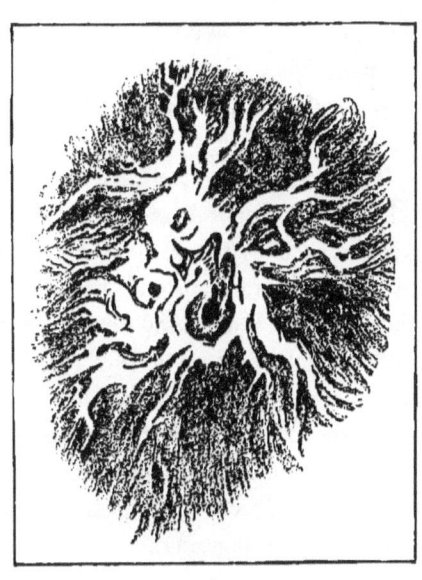

FIG. 28.—FACULA. (FROM A DRAWING BY CHACORNAC.)

This darkening toward the edge, then, means that the sun has an atmosphere which tempers its heat to us. Whatever the sun's heat supply is within its globe, if this atmosphere grow thicker, the heat is more confined within, and our earth will grow colder; if the solar atmosphere grow thinner, the sun's energy will be expended more rapidly, and our earth will grow hotter. This atmosphere, then, is in considerable part, at least, the subject of the action of the spots; this is what they are supposed to carry down or to spout up.

We shall return to the study of it again; but what I want to point out now is that the temperature of the earth, and even the existence of man upon it, depends very much upon this, at first sight, insignificant phenomenon. What, then, is the solar atmosphere? Is it a permanent thing? Not at all. It is more light and unsubstantial than our own air, and is being whirled about by solar winds as ours toss the dust of the streets. It is being sucked down within the body of the sun by some action we do not clearly understand, and returned to the surface by some counter effect which we comprehend no better; and upon this imperfectly understood exchange depends in some way our own safety.

There used to be recorded in medical books the case of a boy who, to represent Phœbus in a Roman mask, was gilded all over to produce the effect of the golden-rayed god, but who died in a few hours because, all the pores of the skin being closed by the gold-leaf, the natural circulation was arrested. We can count with the telescope millions of pores upon the sun's surface, which are in some way connected with the interchange which has just been spoken of; and if this, his own natural circulation, were arrested or notably diminished, we should see his face grow cold, and know that our own health, with the life of all the human race, was waiting on his recovery.

II.

THE SUN'S SURROUNDINGS.

AS I write this, the fields glitter with snow-crystals in the winter noon, and the eye is dazzled with a reflection of the splendor which the sun pours so fully into every nook that by it alone we appear to see everything.

Yet, as the day declines, and the glow of the sunset spreads up to the zenith, there comes out in it the white-shining evening star, which not the light, but the darkness, makes visible; and as the last ruddy twilight fades, not only this neighbor-world, whose light is fed from the sunken sun, but other stars appear, themselves self-shining suns, which were above us all through the day, unseen because of the very light.

As night draws on, we may see the occasional flash of a shooting-star, or perhaps the auroral streamers spreading over the heavens; and remembering that these will fade as the sun rises, and that the nearer they are to it the more completely they will be blotted out, we infer that if the sun were surrounded by a halo of only similar brightness, this would remain forever invisible, — unless, indeed, there were some way of cutting off the light from the sun without obscuring its surroundings. But if we try the experiment of holding up a screen which just conceals the sun, nothing new is seen in its vicinity, for we are also lighted by the neighboring sky, which is so dazzlingly bright with reflected light as effectually to hide anything which may be behind it, so that to get rid of this glare we should need to hang up a screen *outside* the earth's atmosphere altogether.

Nature hangs such a screen in front of the earth when the moon passes between it and the sun; but as the moon is far too small to screen all the earth completely, and as so limited a portion of its surface is in complete shadow that the chances are much against any given individual's being on the single spot covered by it, many centuries usually elapse before such a *total* eclipse occurs at any given point; while yet almost every year

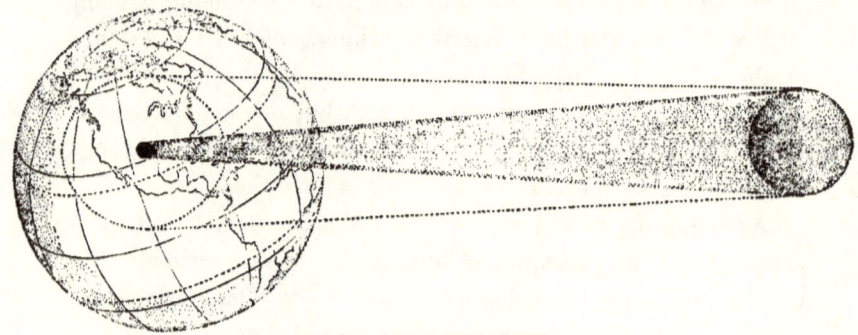

FIG. 29. — LUNAR CONE SHADOW.

there may be a partial eclipse, when, over a great portion of the earth at once, people may be able to look round the moon's edge and see the sunlight but partly cut off. Nearly every one, then, has seen a partial eclipse of the sun, but comparatively few a total one, which is quite another thing, and worth a journey round the world to behold; for such a nimbus, or glory, as we have suggested the possibility of, does actually exist about the sun, and becomes visible to the naked eye on the rare occasions when it is visible at all, accompanied by phenomena which are unique among celestial wonders.

The "corona," as this solar crown is called, is seen during a total eclipse to consist of a bright inner light next the invisible sun, which melts into a fainter and immensely extended radiance (the writer has followed the latter to the distance of about ten million miles), and all this inner corona is filled with curious detail. All this is to be distinguished from another remarkable feature seen at the same time; for close to the black body of the moon are prominences of a vivid crimson and scarlet, rising up like mountains from the hidden solar disk, and these, which will be considered later, are quite distinct from the corona, though seen on the background of its pearly light.

To understand what the lunar screen is doing for us, we may imagine ourselves at some station outside the earth, whence we should behold the moon's shadow somewhat as in Fig. 29, where we must remember that since the lunar orbit is not a circle, but nearly an ellipse, the moon is at some times farther from the earth than at others. Here the extremity of its shadow is represented as just touching the surface of the globe, while it is evident that if the moon were at its greatest distance, its shadow might come to a point before reaching the earth at all. We speak, of course, only of the central cone of shade; for there is an outer one, indicated by the faint dotted lines, within whose much more extended limits the eclipse is partial, but with the latter we have at present nothing to do. The figure however, for want of room, is made to represent the proportions incorrectly, the real ones of the shadow being actually something like those of a sewing-needle,—this very long attenuated shadow sometimes, as we have just said, not reaching the earth at all, and when it does reach it, covering at the most a very small region indeed. Where this point touches, and wherever it rests, we should, in looking down from our celestial station, see that part of the earth in complete shadow, appearing like a minute dark spot, whose lesser diameter is seldom over a hundred and fifty miles.

The eclipse is total only to those inhabitants of the earth within the track of this dark spot, though the spot itself travels across the earth with the speed of the moon in the sky; so that if it could leave a mark, it would in a few hours trace a dark line across the globe, looking like a narrow black tape curving across the side of the world next the sun. In Fig. 30, for instance, is the central track of the eclipse of July 29, 1878, as it would be visible to our celestial observer, beginning in Alaska in the forenoon, and ending in the Gulf of Mexico, which it reached in the afternoon. To those on the earth's surface within this shadow it covered everything in view, and, for anything those involved in it could see, it was all-embracing and terrible, and worthily described in such lines as Milton's, —

> "As when the sun . . .
> In dim eclipse, disastrous twilight sheds
> On half the nations, and with fear of change
> Perplexes monarchs."

We may enjoy the poet's vision; but here, while we look down on the whole earth at once, we must admit that the actual area of the "twilight" is very small indeed. Within this area, however, the spectacle is one of which, though the man of science may prosaically state the facts, perhaps only the poet could render the impression.

We can faintly picture, perhaps, how it would seem, from a station near the lunar orbit, to see the moon — a moving world — rush by with a velocity greater than that of the cannon-ball in its swiftest flight: but with equal speed its shadow actually travels along the earth. And now, if we return from our imaginary station to a real one here below, we are better prepared to see why this flying shadow is such a unique spectacle; for, small as it may be when seen in relation to the whole globe, it is immense to the observer, whose entire horizon is filled with it, and who sees the actual velocity of one of the heavenly bodies, as it were, brought down to him.

The reader who has ever ascended to the Superga, at Turin, will recall the magnificent view, and be able to understand the good fortune of an observer (Forbes) who once had the opportunity to witness thence this phenomenon, and under a nearly cloudless sky. "I perceived," he says, "in the southwest a black shadow like that of a storm about to break, which obscured the Alps. It was the lunar shadow coming toward us." And he speaks of the "stupefaction"—it is his word—caused by the spectacle. "I confess," he continues, "it was the most terrifying sight I ever saw. As always happens in the cases of sudden, silent, unexpected movements, the spectator confounds real and relative motion. I felt almost giddy for a moment, as though the massive building under me bowed on the side of the coming eclipse." Another witness, who had been looking at some bright clouds just before, says: "The bright cloud I saw distinctly put out like a candle. The rapidity of the shadow, and the intensity, produced a feeling that something material was sweeping over the earth at a speed perfectly frightful. I involuntarily listened for the rushing noise of a mighty wind."

FIG. 30.—TRACK OF LUNAR SHADOW.

Each one notes something different from another at such a time; and though the reader will find minute descriptions of the phenomena already in print, it will perhaps be more interesting if, instead of citations from books, I invite him to view them with me, since each can tell best what he has personally seen.

I have witnessed three total eclipses, but I do not find that repetition dulls the interest. The first was that of 1869, which passed across the United States and was nearly central over Louisville. My station was on the southern border of the eclipse track, not very far from the Mammoth Cave in Kentucky, and I well remember that early experience. The special observations of precision in which I was engaged would not interest the reader; but while trying to give my undivided attention to these, a mental photograph of the whole spectacle seemed to be taking without my volition. First, the black body of the moon advanced slowly on the sun, as we have all seen it do in partial eclipses, without anything noticeable appearing; nor till the sun was very nearly covered did the light of day about us seem much diminished. But when the sun's face was reduced to a very narrow crescent, the change was sudden and startling, for the light which fell on us not only dwindled rapidly, but became of a kind unknown before, so that a pallid appearance overspread the face of the earth with an ugly livid hue; and as this strange wanness increased, a cold seemed to come with it. The impression was of something *unnatural;* but there was only a moment to note it, for the sun went out as suddenly as a blown-out gas-jet, and I became as suddenly aware that all around, where it had been, there had been growing into vision a kind of ghostly radiance, composed of separate pearly beams, looking distinct each from each, as though the black circle where the sun once was, bristled

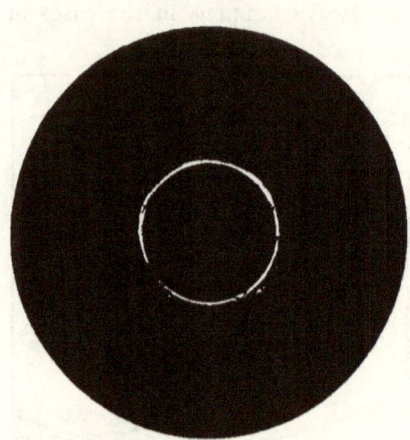

FIG. 31.—INNER CORONA ECLIPSE OF 1869. FROM SHELBYVILLE PHOTOGRAPH. (ROYAL ASTRONOMICAL SOCIETY'S MEMOIRS.)

with pale streamers, stretching far away from it in a sort of crown.

This was the mysterious corona, only seen during the brief moments while the shadow is flying overhead; but as I am undertaking to recall faithfully the impressions of the instant, I may admit that I was at the time equally struck with a circumstance that may appear trivial in description, — the extraordinary globular appearance of the moon herself. We all know well enough that the moon is a solid sphere, but it commonly *looks* like a bright, flat circle fastened to the concave of the starry vault; and now, owing to its unwonted illumination, the actual rotundity was seen for the first time, and the result was to show it as it really is, — a monstrous, solid globe, suspended by some invisible support above the earth, with nothing apparent to keep it from tumbling on us, looking at the moment very near, and more than anything else like a gigantic black cannon-ball, hung by some miracle in the air above the neighboring cornfield. But in a few seconds all was over; the sunlight flashed from one point of the moon's edge and then another, almost simultaneously, like suddenly kindled electric lights, which as instantly flowed into one, and it was day again.

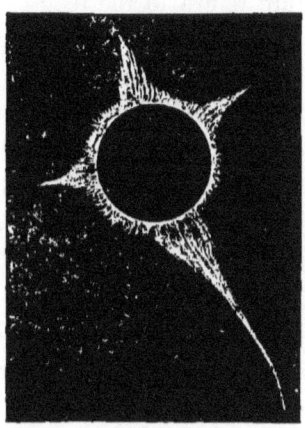

FIG. 32. — SKETCH OF OUTER CORONA, 1869. (U. S. COAST SURVEY REPORT.)

I have spoken of the "unnatural" appearance of the light just before totality. This is not due to excited fancy, for there is something so essentially different from the natural darkness of twilight, that the brute creation shares the feeling with us. Arago, for instance, mentions that in the eclipse of 1842, at

Perpignan, where he was stationed, a dog which had been kept from food twenty-four hours was, to test this, thrown some bread just before "totality" began. The dog seized the loaf, began to devour it ravenously, and then, as the appearance already described came on, he dropped it. The darkness lasted some minutes, but not till the sun came forth again did the poor creature return to the food. It is no wonder, then, that men also, whether educated or ignorant, do not escape the impression. A party of the courtiers of Louis XV. is said to have gathered round Cassini to witness an eclipse from the terrace of the Paris observatory, and to have been laughing at the populace, whose cries were heard as the light began to fade; when, as the unnatural gloom came quickly on, a sudden silence fell on them too, the panic terror striking through their laughter. Something common to man and the brute speaks at such times, if never before or again; something which is not altogether physical apprehension, but more like the moral dismay when the shock of an earthquake is felt for the first time, and we first know that startling doubt, superior to reason, whether the solid frame of earth is real, and not "baseless as the fabric of a vision."

But this is appealing for illustration to an experience which most readers have doubtless been spared,[1] and I would rather cite the lighter one of our central party that day, a few miles north of me, at Shelbyville. In this part of Kentucky the colored population was large, and (in those days) ignorant of everything outside the life of the plantation, from which they had only lately been emancipated. On that eventful 8th of August they came in great numbers to view the enclosure and the tents of the observing party, and to inquire the price of the show. On learning that they might see it without charge from the outside, a most unfavorable opinion was created among them as to the probable merits of so cheap a spectacle, and

---

[1] This was written before the "Charleston earthquake" occurred.

they crowded the trees about the camp, shouting to each other sarcastic comments on the inferior interest of the entertainment. "Those trees there," said one of the observers to me the next day, "were black with them, and they kept up their noise till near the last, when they suddenly stopped, and all at once, and as 'totality' came, we heard a wail and a noise of tumbling, as

FIG. 33.—TACCHINI'S DRAWING OF CORONA OF 1870.
(SECCHI'S "LE SOLEIL.")

though the trees had been shaken of their fruit, and then the boldest did not feel safe till he was under his own bed in his own cabin."

It is impossible to give an exact view of what our friends at Shelbyville saw, for no drawings made there appear to have been preserved, and photography at that time could only indicate feebly the portion of the corona near the sun where it is brightest. Fig. 31 is a fac-simile of one of the photographs taken on the occasion, which is interesting perhaps as one of the early attempts in this direction, for comparison with later ones; but as a picture it is very disappointing, for the whole structure of the outer corona we have alluded to is missed altogether, the plate having taken no impression of it.

A drawing (Fig. 32) made by another observer, Mr. M'Leod, at Springfield, represents more of the outer structure; but the reader must remember that all drawings must, in the nature of the case (since there are but two or three minutes to sketch in), be incomplete, whatever the artist's skill.

Up to this time it was still doubtful, not only what the corona was, but where it was; whether it was a something about the sun or moon, or whether, indeed, it might not be in our own atmosphere. The spectroscopic observations of Professors Young and Harkness at this same eclipse of a green line in its spectrum, due to some glowing gas, showed conclusively that it was largely, at any rate, a solar appendage, and partly, at least, self-luminous; and these and other results having awakened general discussion among astronomers in Europe as well as at home, the United States Government sent an expedition, under the direction of the late Professor Pierce, to observe an eclipse which in the next year, on Dec. 8, 1870, was total in the south of Spain. There were three parties; and of the most western of these, which was at Xeres under the charge of Professor Winlock, I was a member.

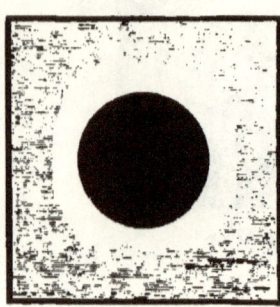

FIG. 34. — WATSON'S NAKED-EYE DRAWING OF CORONA OF 1870. (U. S. COAST SURVEY REPORT.)

The duration of totality was known beforehand. It would last two minutes and ten seconds, and to secure what could be seen in this brief interval we crossed the ocean. Our station was in the midst of the sherry district, and a part of the instruments were in an orange-grove, where the ground was covered

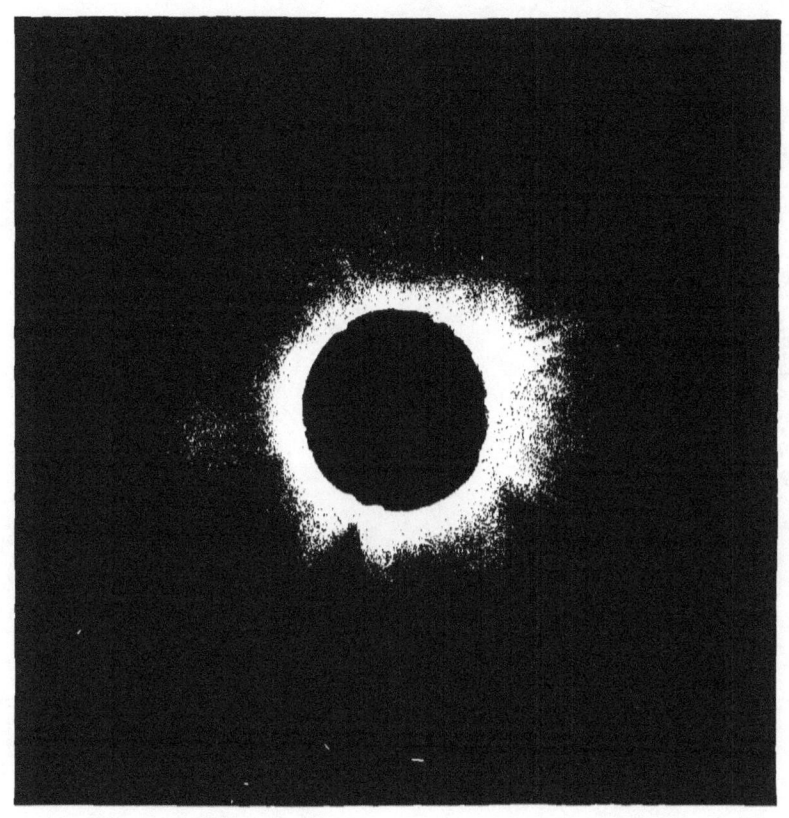

FIG. 35.—PHOTOGRAPH SHOWING COMMENCEMENT OF OUTER CORONA.
(ROYAL ASTRONOMICAL SOCIETY'S MEMOIRS.)

with the ripe fallen fruit, while the olive and vine about us in December reminded us of the distance we had come to gather the results of so brief an opportunity.

To prepare for it, we had all arrived on the ground some weeks beforehand, and had been assiduously busy in installing the apparatus in the observing camp, which suggested that of a small army, the numerous instruments, some of them of considerable size, — equatorials, photographic apparatus, polariscopes, photometers, and spectroscopes, — being under tents, the fronts of which could be lifted when the time came for action.

To the equatorial telescopes photographic cameras are attached instead of the eye-pieces, in the hope that the corona may be made to impress itself on the plate instead of on the eye. The eye is an admirable instrument itself, no doubt; but behind it is a brain, perhaps overwrought with excitement, and responding too completely to the nervous tension which most of us experience when those critical moments are passing so rapidly. The camera can see far less of the corona than the man, *but it has no nerves*, and what it sets down we may rely on.

At such a time each observer has some particular task assigned to him, on which, if wise, he has drilled himself for weeks beforehand, so that no hesitation or doubt may arise in the moment of action; and his attention is expected to be devoted to this duty alone, which may keep him from noting any of the features which make the occasion so impressive as a spectacle. Most of my own particular work was again of a kind which would not interest the reader.

Apart from this, I can recall little but the sort of pain of expectation, as the moment approached, till within a minute before totality the hum of voices around ceased, and an utter and most impressive silence succeeded, broken only by a low "Ah!" from the group without the camp, when the moment came. I remember that the clouds, which had hung over the sun while the moon was first advancing on its body, cleared away before the instant of totality, so that the last thing I saw was a range of mountains to the eastward still bright in the light; then, the next moment, the shadow rushed overhead and blotted out the distant hills, almost before I could turn my face to the instrument before me.

FIG. 36. — ECLIPSE OF 1857, DRAWING BY LIAIS. (ROYAL ASTRONOMICAL SOCIETY'S MEMOIRS.)

The corona appeared to me a different thing from what it did the year before. It was apparently confined to a pearly light of a roughly quadrangular shape, close to the limb of the sun, broken by dark rifts (one of which was a conspicuous object); while within, and close to the limb, was what looked like a mountain rising from the hidden sun, of the color of the richest tint we should see in a rose-leaf held up against the light, while others were visible of an orange-scarlet. After a short scrutiny I turned to my task of analyzing the nature of the white light.

The seconds fled, the light broke out again, and so did the hubbub of voices, — it was all over, and what had been missed then could not be recovered. The sense of self-reproach for wasted opportunity is a common enough feeling at this time, though one may have done his best, so little it seems to each he has accomplished; but when all the results had been brought together, we found that the spectroscopes, cameras, and polariscopes had each done their work, and the journey had not been taken in vain. In one point only we all differed, and this was about the direct ocular evidence, for each seemed to have seen a different corona, and the drawings of it were singularly unlike. Here are two (Figs. 33 and 34) taken at this eclipse at the same time, and from neighboring stations, by two most experienced astronomers, Tacchini and Watson. No one could guess that they represented the same object, and a similar discrepancy was common.

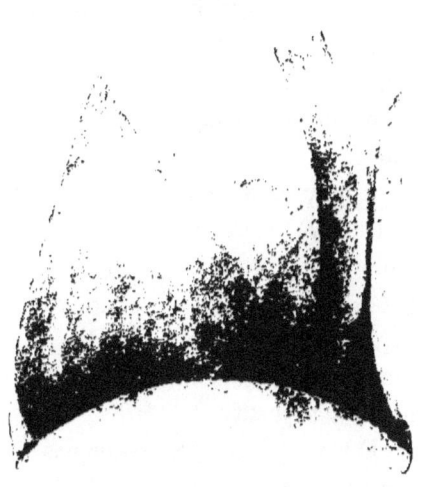

FIG. 37.— ENLARGEMENT OF PART OF FIG. 38.

Considering that these were trained experts, whose special task it was, in this case, to draw the corona, which therefore claimed their undivided attention, I hardly know a more striking instance of the fallibility of human testimony. The evidence of several observers, however, pointed to the fact that the light really was more nearly confined to the part next the sun than the year before, so that the corona had probably changed during that interval, and grown smaller, which was remarkable enough. The evidence of the polariscope, on the whole, showed it to be partly due to reflected sunlight, while the spectroscope in the hands of Professor Young confirmed the last year's observation, that it was also, and largely, self-luminous. Finally, the photographs, taken at very distant stations, showed the same dark rifts in the same place, and thus brought confirmatory evidence that it was not a local phenomenon in our own atmosphere. A photograph of it, taken by Mr. Brothers in Sicily, is the subject of the annexed illustration (Fig. 35), in which the very bright lights which, owing to "photographic irradiation," seem to indent the moon, are chiefly due to the colored flames I have spoken of, which will be described later.

It may be observed that the photographs taken in the next year (1871) were still more successful, and began to show still more of the structure, whose curious forms, resembling large petals, had already been figured by Liais. His drawing (Fig. 36), made in 1857, was supposed to be rather a fanciful sketch than a trustworthy one; but, as it will be seen, the photograph goes far to justify it.

Figures 37 and 38 are copies published by Mr. Ranyard of the excellent photographs obtained in 1871, which are perhaps as good as anything done since, though even these do not show the outer corona. The first is an enlargement of a small portion of the detail in the second. It is scarcely possible for wood-engraving to reproduce the delicate texture of the original.

The years brought round the eclipse of 1878, which was again in United States territory, the central track (as Fig. 30 has already shown) running directly over one of the loftiest mountains of the country, Pike's Peak, in Colorado. Pike's Peak, though over fourteen thousand feet high, is often ascended by pleasure tourists; but it is one thing to stay there for an hour or two, and another to take up one's abode there and get acclimated, — for to do the latter we must first pass through the horrors (not too strong a word) of mountain-sickness. This

FIG. 38.—FAC-SIMILE OF PHOTOGRAPH OF CORONA OF 1871.
(ROYAL ASTRONOMICAL SOCIETY'S MEMOIRS.)

reaches its height usually on the second or third day, and is something like violent sea-sickness, complicated with the sensations a mouse may be supposed to have under the bell of an air-pump. After a week the strong begin to get over it, but none but the very robust should take its chances, as we did, without preparation; for on the night before the eclipse the life of one of our little party was pronounced in danger, and he was carried down in a litter to a cabin at an altitude of about ten thousand feet, where he recovered so speedily as to be able to do good service on the following day. The summit of the "Peak" is covered with great angular bowlders of splintered granite, among which we laid logs brought up for firewood, and on these, sacks of damp hay, then stretching a little tent over all and tying it down with wire to the rocks, we were fain to turn in under damp blankets, and to lie awake with incessant headache, drawing long, struggling breaths in the vain attempt to get air, and wondering how long the tent would last, as the canvas flapped and roared with a noise like that of a loose sail in a gale at sea, with occasional intervals of a dead silence, usually followed by a gust that shoved against the tent with the push of a solid body, and if a sleeper's shoulders touched the canvas, shouldered him over in his bed. The stout canvas held, but the snow entered with the wind and lay in a deep drift on the pillow, when I woke after a brief sleep toward morning, and, looking out on the gray dawn, found that the snow had turned to hail, which was rattling sharply on the rocks with an accompaniment of thunder, which seemed to roll from all parts of the horizon. The snow lay thick, and the sheets of hail were like a wall, shutting out the sight of everything a few rods off, and this was in July! I thought of my December station in sunny Andalusia.

Hail, rain, sleet, snow, fog, and every form of bad weather continued for a week on the summit, while it was almost always clear below. It was often a remarkable sight to go to the edge and look down. The expanse of "the plains," which stretched eastward to a horizon line over a hundred miles distant, would be in bright sunshine beneath, while the hail was all around and above us; and the light coming *up* instead of down gave sin-

FIG. 39. — "SPECTRES."

gular effects when the clouds parted below, the plains seeming at such times to be opalescent with luminous yellow and green, as though the lower world were translucent, and the sun were beneath it and shining up through. Fig. 39 is a picture of three of us on the mountain-top, who saw a rarer spectacle; for directly opposite the setting sun, and on the mist over the gulf beyond us, was a bright ring, in whose centre were three phantom images of our three selves, which moved as we moved, and then faded as the sun sank. It was "the spectre of the Brocken." These ghostly presentments were tolerably defined, as in the sketch, but did not seem to be gigantic, as some have described them. We rather thought them close at hand; but before we could determine, the vision faded.

The clouds, to our good fortune, rolled away on the 29th; and a number of pleasure-seekers, who came up to view the eclipse and the unwonted bright sunshine, made a scene which it was hard to identify with the usual one. This time my business was to draw the corona; and the extreme altitude and the clearness of the air, with perhaps some greater extension than usual in the object itself, enabled it to be followed to an unprecedented distance. During totality the sun was surrounded by a narrow ring — hardly more than a line — of vivid light, presenting no structure to the naked eye (but a remarkable one in the telescope); and this faded with great suddenness into a circular nebulous luminosity between two and three diameters of the sun wide, but without such marked plumes, or filaments, as I had seen in 1869. The most extraordinary thing, however, was a beam of light, inclined at an angle of about forty-five degrees, about as wide as the sun, and extending to the distance of nearly six of its diameters on one side and over twelve on the other; on one side alone, that is, to the amazing distance of over ten million miles from its body. Substantially the same observation was made, as it appeared later, by Professor Newcomb, at a lower level. The direction, when more carefully measured, it was interesting to note, coincided closely with that of the Zodiacal light, and a faint central rib added to its resemblance to that body. It is noteworthy, in illustration of what has already been said as to the conflict of ocular testimony, that though I, with the great majority of observers below, saw only this beam, two witnesses whose evidence is unimpeachable, Professors Young and Abbe, saw a pale beam at right angles to it; and that one observer did not see the beam in question at all. Fig. 40 is a sketch made from my own, but necessarily on a scale which can show only its general features.

With the telescope, the whole of the bright inner light close to the sun was found to be made up of filaments, more definite even than those described in a previous chapter as seen in sunspots, and bristling in all directions from the edge; not concealing each other, as we might expect such things to do, upon a sphere, but fringing the sun's edge in definite outline, as though it were really but a disk.

Those who were at leisure to watch the coming shadow of the moon described its curved outline as distinctly visible on the plains. "A rounded ball of darkness with an orange-yellow border," one called it. Those, again, who looked down on the bright clouds below say the shadow was preceded by a yellow fringe, casting a bright light over the clouds and passing into orange, pink, rose-red, and dark-red, in about twenty seconds. This beautiful effect was noticed by nearly all the amateur observers present, who had their attention at liberty, and was generally unseen by the professional ones, who were shut up in dark tents with photometers, or engaged otherwise than in admiring the glory of the spectacle as a spectacle merely. This strange light, forming a band of color about the shadow as seen from above, must have really covered ten miles or more in width, and have occupied a considerable fraction of a minute in passing over the heads of those below, to whom it probably constituted that lurid light on their landscape I have spoken of as so peculiar and "unnatural." It seems to be due to the colored flames round the sun, which shine out when its brighter light is extinguished. I should add that on the summit of Pike's Peak the corona did not entirely disappear at the instant the sun

FIG. 40.—OUTER CORONA OF 1878. (U. S. NAVAL OBSERVATORY.)

broke forth again, but that its outlying portions first went and then its brighter and inner ones, till our eager gaze, trying to follow it as long as possible, only after the lapse of some minutes saw the last of the wonderful thing disappear and "fade into the light of common day."

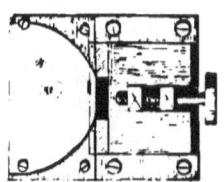

FIG. 41.—SPECTROSCOPE SLIT AND SOLAR IMAGE. (FROM "THE SUN," BY YOUNG.)

There have been other eclipses since; but, in spite of all, our knowledge of the corona remains very incomplete, and if the most learned in such matters were asked what it was, he could probably answer truthfully, "I don't know."

This will not be wondered at when it is considered that as total eclipses come about every other year, and continue, one with another, hardly three minutes, an astronomer who should devote thirty years exclusively to the subject, never missing an eclipse in whatever quarter of the globe it occurred, would in that time have secured, in all, something like three-quarters of an hour for observation. Accordingly, what we know best about the corona is how it looks, what it *is* being still largely conjecture; and it is for this reason that I have thought the space devoted to it would be best used by giving the unscientific reader some idea of the visible phenomena as they present themselves to an eyewitness. Treatises like Lockyer's "Solar Physics," Proctor's "The Sun," Secchi's "Le Soleil," and Young's "The Sun" (the latter is most recent), will give the reader who desires to learn more of the little that is known, the fuller infor-

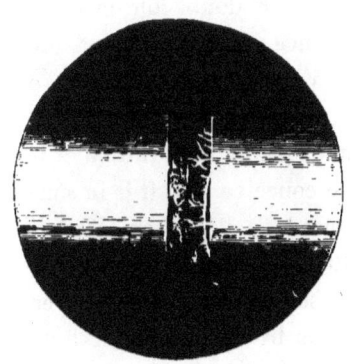

FIG. 42.—SLIT AND PROMINENCES. ("THE SUN," BY YOUNG.)

mation which this is not the place for; but it may be said very briefly that it is certain that the corona is at times of enormous extent (the whole length of the longer beam seen on Pike's Peak must have been over fourteen million miles), that it almost certainly changes in its shape and dimensions from year to year (possibly much oftener, but this we cannot yet know), and that it shines partly by its own and partly by reflected light. When we come to ask whether it is a gas or not, the evidence is conflicting. The appearance of the green coronal line, and other testimony we have not alluded to, would make it seem almost certain that there must be a gas here of extreme tenuity, reaching the height of some hundred thousand miles, at the least; while yet the fact that such light bodies as comets have been known to pass through it, close to the sun, without suffering any visible retardation, such as would come even from a gas far lighter than hydrogen, appears to throw doubt on evidence otherwise strong. It is possible to conceive of the corona, and especially of the outer portion, as very largely made up of minute particles such as form the scattered dust of meteoric trains, and this seems to be the most probable constitution of its outlying parts. It is even possible to conceive that it is in some degree a subjective phenomenon, caused, as Professor Hastings has suggested, by diffraction upon the edge of the moon, — the moon, that is, not merely serving as a screen to the sun to reveal the corona, but partly *making* the corona by diffracting the light, somewhat as we see that the edge of any very distant object screening the sun is gilded by its beams. This effect may be seen when the sun rises or sets unusually clear, for objects on the horizon partly hiding it are then fringed for a moment with a line of light, — an appearance which has not escaped Shakspeare, where he says, —

> "But when from under this terrestrial ball
> He fires the tall tops of the eastern pines."

Still, in admitting the possibility of some such contributory effect on the part of the moon, we must not, of course, be understood as meaning that the corona as a whole does not have a real existence, quite independent of the changes which the presence of the moon may bring; and in leaving the wonderful thing we must remember that it is, after all, a reality, and not a phantasm.

I have already described how, at the eclipse of 1870, I (with others) saw within the corona what seemed like rose and scarlet-colored mountains rising from the sun's edge, an appearance which had first been particularly studied in the eclipse of 1868, two years before, and which, it might be added, Messrs. Lockyer and Janssen had succeeded in observing without an eclipse by the spectroscope. Besides the corona, it may be said, then, that the sun is surrounded by a thin envelope, rising here and there into prominences of a rose and scarlet color, invisible in the telescope, except at a total eclipse, but always visible through the spectroscope. It is within and quite distinct from the corona, and is usually called the "chromosphere," being a sort of sphere of colored fire surrounding the sun, but which we can usually see only on the edge. "The appearance," says Young, "is as if countless jets of heated gas were issuing through vents and spiracles over the whole surface, thus clothing it with flame, which heaves and tosses like the blaze of a conflagration." Out of this, then, somewhat like greater waves or larger swellings of the colored fires, rise the prominences, whose place, close to the sun's edge, has been indicated in many of the drawings and photographs just given of the corona, on whose background they are seen during eclipses; but as they can be studied at our leisure with the spectroscope, we have reserved a more particular description of them till now. They are at all times directly before us, as well as the corona; but while both are yet invisible from the overpowering brightness of the sunlight reflected from the earth's atmosphere in front of them, these red flames are so far brighter than the coronal background, that if we could only weaken this "glare" a little, they at least might become visible,

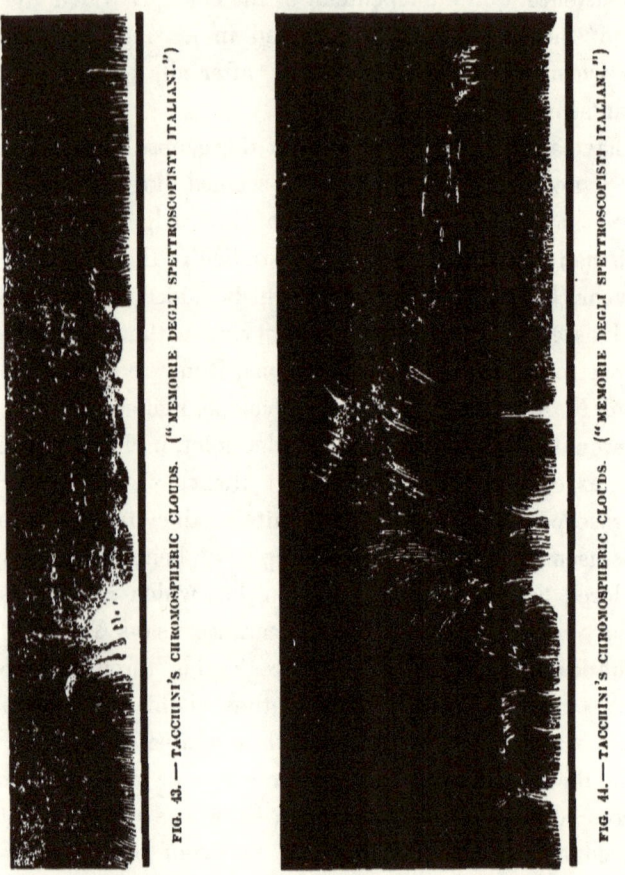

FIG. 43.—TACCHINI'S CHROMOSPHERIC CLOUDS. ("MEMORIE DEGLI SPETTROSCOPISTI ITALIANI.")

FIG. 44.—TACCHINI'S CHROMOSPHERIC CLOUDS. ("MEMORIE DEGLI SPETTROSCOPISTI ITALIANI.")

even if the corona were not. The difficulty is evidently to find some contrivance which will weaken the "glare" without enfeebling the prominences too; and this the spectroscope does by diffusing the white sunlight, while it lets the color pass nearly unimpaired. For the full understanding of its action the reader must be referred to such works as those on the sun already mentioned; but a general idea of it may be gathered, if we reflect that white light is composed of every possible variety of colors, and that the spectroscope, which consists essentially of a prism behind a very narrow slit through which the light enters, lets any single color pass freely, without weakening it or altering it in anything but its direction, but gives a different direction to each, and hence sorts out the tints, distributing them side by side, every one in its own place, upon the long colored band called the spectrum. If this distribution has spread the colors along a space a thousand times as wide as the original beam, the average light must be just so much weaker than the white light was, because this originally consisted of a thousand (let us say a thousand, but it is really an infinite number) mingled tints of blue, green, yellow, orange, and red, which have now been thus distributed. If, however, we look through the prism at a rose-leaf, and it has no blue, green, yellow, or orange in it, and nothing but pure red, as each single color passes unchanged, this red will, according to what has been said, be as bright after it has passed as before. All depends, then, on the fact that these prominences do consist mainly of light of one color, like the rose-leaf, so that this monochromatic light will be seen through the spectroscope just as it is, while the luminous veil of glaring white before it will seem to be brushed away.

If a large telescope be directed toward the sun, the glass at the farther end will, if we remove the eye-piece, form a little picture of the sun, as a picture is formed in a camera-obscura; and now, if we also fasten the spectroscope to this eye-end, where the observer's head would be were he looking through, the edge of the solar image may be made to fall just *off* the slit, so that only the light from the prominences (and the white glare about them) shall pass in. To see this more clearly, let us turn our backs to the sun and the telescope, and look at the place where the image falls by the spectroscope slit, which in Fig. 41 is drawn of its full size. This is a brass plate, having a minute rectangular window, the "slit," in it. The width of this slit is regulated by a screw, and any rays falling into the narrow aperture pass through the prism within, and finally fall on the observer's eye, but not till they have been sorted by the prism in the manner described. Formed on the brass plate, just as it

FIG. 45. — VOGEL'S CHROMOSPHERIC FORMS. ("BEOBACHTUNGEN," DR. H. C. VOGEL.)

would be formed on a sheet of paper, or anything else held in the focus, we see the bright solar image, a circle of light perhaps an inch and a half in diameter, — a miniature of the sun with its spots. The whole of the sun (the photosphere) then is hidden to an observer who is looking up through the slit from the other side, for, as the sun's edge does not quite touch the slit, none of its rays can enter it; but if there be also the image here of a prominence, projecting beyond the edge, and really overhanging the slit (though to us invisible on account of the glare about it), these rays will fall into the slit and pass down to the prism, which will dispose of it in the way already stated.

And now let us get to the other side, and, looking up through the prism with the aid of a magnifying-glass, see what it has done for us (Fig. 42). The large rectangular opening here is the same as the small one which was visible from the outside, only that it is now magnified, and what was before invisible is seen; the edge of the sun itself is just hidden, but the scarlet flames of the chromosphere have become visible, with a cloudy prominence rising above them. The "flames" are flame-like only in form, for their light is probably due not to any combustion, but to the glow of intensely heated matter; and as its light is not quite pure red, we can, by going to another part of the spectrum, see the same thing repeated in orange, the effect being as though we had a number of long narrow windows, some glazed with red, some with orange, and some with other colors, through which we could look out at the same clouds. I have looked at these prominences often in this way; but I prefer, in the reader's interest, to borrow from the description by Professor Young, who has made these most interesting and wonderful forms a special study.

Let us premise that the depth of the crimson shell out of which they rise is usually less than five thousand miles, and that though the prominences vary greatly, the majority reach a height of nearly twenty thousand miles, while in exceptional cases this is immensely exceeded. Professor Young has seen one which grew to a height of three hundred and fifty thousand miles in an hour and a half, and in half an hour more had faded away.

These forms fall into two main classes, — that of the quiet and cloud-like, and that of the eruptive, — the first being almost exactly in form like the clouds of our own sky, sometimes appearing to lie on the limb of the sun like a bank of clouds on the horizon, sometimes floating entirely free; while sometimes "the whole under surface is fringed with down-hanging filaments, which remind one of a summer shower hanging from a heavy thunder-cloud."

Here are some of the typical forms of the quieter ones: —

Fig. 43, by Tacchini, the Director of the Roman Observatory, represents an ordinary prominence, or cloud-group in the chromosphere, whose height is about twenty-five thousand miles. The little spires of flame which rise, thick as grass-blades, everywhere from the surface, are seen on its right and left.

FIG. 46.— TACCHINI'S CHROMOSPHERIC FORMS. ("MEMORIE DEGLI SPETTROSCOPISTI ITALIANI.")

Fig. 44 (Tacchini) is one where the agitation is greater and the "filamentary" type is more marked. Besides the curiously threadlike forms (so suggestive of what we have already seen in the photosphere), we have here what looks like an extended cloudy mass, drawn out by a horizontally moving wind.

Fig. 45 (by Vogel, at Bothkamp) represents another of these numerous types.

The extraordinary Fig. 46 is from another drawing, by Tacchini, of a protuberance seen in 1871 (a time of great solar disturbance), and it belongs to the more energetic of its class.

FIG. 47. — ERUPTIVE PROMINENCES. ("THE SUN," BY YOUNG.)

This fantastic cloud-shape, "if shape it might be called that shape had none," looking like some nightmare vision, was about fifty thousand miles long and sixty thousand high above the surface. The reader will notice also the fiery rain, like the drops from a falling rocket, and may add to it all, in imagination, the actual color, which is of a deep scarlet.

It may add to the interest such things excite, to know that they have some mysterious connection with a terrestrial phenomenon, — the aurora, — for the northern lights have been again and again noticed to dance in company with these solar displays.

The eruptive prominences are very different in appearance, as will be seen by the next illustration, for which we are indebted to Professor Young.

In Fig. 47 we have a group of most interesting views by him (drawn here on the common scale of seventy-five thousand miles to an inch), illustrating the more eruptive types, of which we will let him speak directly. The first shows a case of the vertical filaments, like those rocket-drops we saw just now in Tacchini's drawing, but here more marked; while the second (on the left side) is a cyclone-form, where the twisted stems suggest what we have seen before in the "bridges" of sun-spots, and below this is another example of filamentary forms.

The upper one, on the right, is the view of a cloud prominence as it appeared at *half-past twelve* o'clock, on Sept. 7, 1871. Below it is the same prominence at *one* o'clock (half an hour later), when it has been shattered by some inconceivable explosion, blowing it into fragments, and driving the hydrogen to a height of two hundred thousand miles. The lowest figure on the right shows another case where inclined jets (of hydrogen) were seen to rise to a height of fifty thousand miles.

Professor Young says of these: —

"Their form and appearance change with great rapidity, so that the motion can almost be seen with the eye. Sometimes they consist of pointed rays, diverging in all directions, like hedgehog-spines. Sometimes they look like flames; sometimes like sheaves of grain; sometimes like whirling water-spouts, capped with a great cloud; occasionally they present most exactly the appearance of jets of liquid fire, rising and falling in graceful parabolas; frequently they carry on their edges spirals like the volutes of an Ionic column; and continually they detach filaments which rise to a great elevation, gradually expanding and growing fainter as they ascend, until the eye loses them. There is no end to the number of curious and interesting appearances which they exhibit under varying circumstances. The velocity of the motions often exceeds a hundred miles a second, and sometimes, though very rarely, reaches two hundred miles."

In the case of the particular phenomenon recorded by Professor Young in the last illustration, Mr. Proctor, however, has calculated that the initial velocity probably exceeded five hundred miles a second, which, except for the resistance experienced by the sun's own atmosphere, would have hurled the ejected matter into space entirely clear of the sun's power to recall it, so that it would never return.

It adds to our interest in these flames to know that they at least are connected with that up-rush of heated matter from the sun's interior, forming a part of the circulation which maintains both the temperature of its surface and that radiation on which all terrestrial life depends. The flames, indeed, add of themselves little to the heat the sun sends us, but they are in this way the outward and visible signs of a constant process within, by which we live; and so far they seem to have a more immediate interest to us, though invisible, than the corona which surrounds them. But we must remember when we lift our eyes to the sun that this latter wonder is really there, whether man sees it or not, and that the cause of its existence is still unknown.

We ask for its "object" perhaps with an unconscious assumption that the whole must have been in some way provided to subserve *our* wants; but there is not as yet the slightest evidence connecting its existence with any human need or purpose, and as yet we have no knowledge that, in this sense, it exists to any "end" at all. "As the thought of man is widened with the process of the suns," let us hope that we shall one day know more.

III.

THE SUN'S ENERGY.

"IT is indeed," says good Bishop Berkeley, "an opinion strangely prevailing amongst men that . . . all sensible objects have an existence . . . distinct from their being perceived by the understanding. But . . . some truths there are, so near and obvious to the mind, that a man need only open his eyes to see them. Such I take this important one to be, namely, that all the choir of heaven and furniture of the earth — in a word, all those bodies which compose the mighty frame of the world — have not any subsistence without a mind."

We are not going to take the reader along "the high priori road" of metaphysics, but only to speak of certain accepted conclusions of modern experimental physics, which do not themselves, indeed, justify all of Berkeley's language, but to which these words of the author of "A New Theory of Vision" seem to be a not unfit prelude.

When we see a rose-leaf, we see with it what we call a color, and we are apt to think it is in the rose. But the color is in *us*, for it is a sensation which something coming from the sun excites in the eye; so that if the rose-leaf were still there, there would be no color unless there were an eye to receive and a brain to interpret the sensation. Every color that is lovely in the rainbow or the flower, every hue that is vivid in a ribbon or sombre in the grave harmonies of some old Persian rug, the metallic lustre of the humming-bird or the sober imperial yellow of precious china, — all these have no existence as color apart from the seeing eye, and all have their fount and origin in the sun itself.

"Color" and "light," then, are not, properly speaking, external things, but names given to the sensations caused by an uncomprehended something radiated from the sun, when this falls on our eyes. If this very same something falls on our face, it produces another kind of sensation, which we call "heat," or if it falls on a thermometer it makes it rise; while if it rests long on the face it will produce yet another effect, "chemical action," for it will *tan* the cheek, producing a chemical change there; or it will do the like work more promptly if it meet a photographic plate. If we bear in mind that it is the identically same thing (whatever that is) which produces all these diverse effects, we see, some of us perhaps for the first time, that "color," "light," "radiant heat," "actinism," etc., are only names given to the diverse effects of some thing, not things themselves; so that, for instance, all the splendor of color in the visible world *exists only in the eye that sees it.* The reader must not suppose that he is here being asked to entertain any metaphysical subtlety. We are considering a fact almost universally accepted within the last few years by physicists, who now generally admit the existence of a something coming from the sun, which is not itself light, heat, or chemical action, but of which these are effects. When we give this unknown thing a name, we call it "radiant energy."

How it crosses the void of space we cannot be properly said to know, but all the phenomena lead us to think it is in the form of motion in some medium, — somewhat (to use an imperfect analogy) like the transmission through the air of the vibrations which will cause sound when they reach an ear. This, at any rate, is certain, that there is an action of some sort incessantly going on between us and the sun, which enables us to experience the effects of light and heat. We assume it to be a particular mode of vibration; but whatever it is, it is repeated with incomprehensible rapidity. Experiments recently made by the writer show that the *slower* heat vibrations which reach us from the sun succeed each other nearly 100,000,000,000,000 times in a single second, while those which make us see, have long been known to be more rapid still. These pass outward from the sun in every direction, in ever-widening spheres; and in them, so far as we know, lies the potency of life for the planet upon whose surface they fall.

Did the reader ever consider that next to the mystery of gravitation, which draws all things on the earth's surface down, comes that mystery — not seen to be one because so familiar — of the occult force in the sunbeams which lifts things *up?* The incomprehensible energy of the sunbeam brought the carbon out of the air, put it together in the weed or the plant, and lifted each tree-trunk above the soil. The soil did not lift it, any more than the soil in Broadway lifted the spire of Trinity. Men brought stones there in wagons to build the church, and the sun brought the materials in its own way, and built up alike the slender shaft that sustains the grass blade and the column of the pine. If the tree or the spire fell, it would require a certain amount of work of men or horses or engines to set it up again. So much actual work, at least, the sun did in the original building; and if we consider the number of trees in the forest, we see that this alone is something great. But besides this, the sun locked up in each tree a store of energy thousands of times greater than that which was spent in merely lifting the trunk from the ground, as we may see by unlocking it again, when we burn the tree under the boiler of an engine; for it will develop a power equal to the lifting of thousands of its kind, if we choose to employ it in this way. This is so true, that the tree may fall, and turn to coal in the soil, and still keep this energy imprisoned in it, — keep it for millions of years, till the black lump under the furnace gives out, in the whirling spindles of the factory or the turning wheel of the steamboat, the energy gathered in the sunshine of the primeval world.

The most active rays in building up plant-life are said to be the yellow and orange, though Nature's fondness for green everywhere is probably justified by some special utility. At any rate, the action of these solar rays is to decompose the products of combustion, to set free the oxygen, and to fix the carbon in the plant. Perhaps these words do not convey a definite meaning to the reader, but it is to be hoped they will, for the statement they imply is wonderful enough. Swift's philosopher at Laputa, who had a project for extracting sunbeams out of cucumbers, was wiser than his author knew; for cucumbers, like other vegetables, are now found to be really in large part put together by sunbeams, and sunbeams, or what is scarcely distinguishable from such, could with our present scientific knowledge be extracted from cucumbers again, only the process would be too expensive to pay. The sunbeam, however, does what our wisest chemistry cannot do: it takes the burned out ashes and makes them anew into green wood; it takes the close and breathed out air, and makes it sweet and fit to breathe by means of the plant, whose food is the same as our poison. With the aid of sunlight a lily would thrive on the deadly atmosphere of the "black hole of Calcutta;" for this bane to us, we repeat, is vital air to the plant, which breathes it in through all its pores, bringing it into contact with the chlorophyl, its green blood, which is to it what the red blood is to us; doing almost everything, however, by means of the sun ray, for if this be lacking, the oxygen is no longer set free or the carbon retained, and the plant dies. This too brief statement must answer instead of a fuller description of how the sun's energy builds up the vegetable world.

But the ox, the sheep, and the lamb feed on the vegetable, and we in turn on them (and on vegetables too); so that, though we might eat our own meals in darkness and still live, the meals themselves are provided literally at the sun's expense, virtue having gone out of him to furnish each morsel we put in our mouths. But while he thus prepares the material for our own bodies, and while it is plain that without him we could not exist any more than the plant, the processes by which he acts grow more intricate and more obscure in our own higher organism, so that science as yet only half guesses how the sun makes us. But the making is done in some way by the sun, and so almost exclusively is every process of life.

It is not generally understood, I think, how literally true this is of every object in the organic world. In a subsequent illustration we shall see a newspaper being printed by power directly and visibly derived from the sunbeam. But all the power derived from coal, and all the power derived from human muscles, comes originally from the sun, in just as literal a sense; for the paper on which the reader's eye rests was not only made primarily from material grown by the sun, but was stitched together by derived sun-power, and by this, also, each page was printed, so that the amount of this solar radiation expended for printing each chapter of this book could be stated with approximate accuracy in figures. To make even the reader's hand which holds this page, or the eye which sees it, energy again went out from the sun; and in saying this I am to be understood in the plain and common meaning of the words.

Did the reader ever happen to be in a great cotton-mill, where many hundreds of operatives watched many thousands of spindles? Nothing is visible to cause the multiplied movement, the engine being perhaps away in altogether another building. Wandering from room to room, where everything is in motion derived from some unseen source, he may be arrested in his walk by a sudden cessation of the hum and bustle, — at once on the floor below, and on that above, and all around him. The simultaneousness of this stoppage at points far apart when the steam is turned off, strikes one with a sense of the intimate dependence of every complex process going on upon some remote invisible motor. The cessation is not, however, absolutely instantaneous; for the great fly-wheel, in which a trifling part of the motor power is stored, makes one or two turns more, till the energy in this also is exhausted, and all is still. The coal-beds and the forests are to the sun what the fly-wheel is to the engine: all their power comes from him; they retain a little of it in store, but very little by comparison with the original; and were the change we have already spoken of to come over the sun's circulation, — were the solar engine disconnected from us, — we could go on perhaps a short time at the cost of this store, but when this was over it would be over with us, and all would be still here too.

Is there not a special interest for us in that New Astronomy which considers these things, and studies the sun, not only in the heavens as a star, but in its workings here, and so largely in its relations to man?

Since, then, we are the children of the sun, and our bodies a product of its rays, as much as the ephemeral insects that its heat hatches from the soil, it is a worthy problem to learn how things earthly depend upon this material ruler of our days. But although we know it does nearly all things done on the earth, and have learned a little of the way it builds up the plant, we know so little of the way it does many other things here that we are still often only able to connect the terrestrial effect with the solar cause by noting what events happen together. We are in this respect in the position of our forefathers, who had not yet learned the science of electricity, but who noted that when a flash of lightning came a clap of thunder followed, and concluded as justly as Franklin or Faraday could have done that there was a physical relation between them. Quite in this way, we who are in a like position with regard to the New Astronomy, which we hope will one day explain to us what is at present mysterious in our connection with the sun, can as yet often only infer that when certain phenomena there are followed or accompanied by others here, all are really connected as products of one cause, however dissimilar they may look, and however little we know what the real connection may be.

There is no more common inquiry than as to the influence of sun-spots on the weather; but as we do not yet know the real nature of the connection, if there be any, we can only try to find out by assembling independent records of sun-spots and of the weather here, and noticing if any changes in the one are accompanied by changes in the other; to see, for instance, if when sun-spots are plenty the weather the world over is rainy or not, or to see if when an unusual disturbance breaks out in a sun-spot any terrestrial disturbance is simultaneously noted.

When we remember how our lives depend on a certain circulation in the sun, of which the spots appear to be special examples, it is of interest not only to study the forms within them, as we have already been doing here, but to ask whether the spots themselves are present as much one year as another. The sun sometimes has numerous spots on it, and sometimes none at all; but it does not seem to have occurred to any one to see whether they had any regular period for coming or going, till Schwabe, a magistrate in a little German town, who happened to have a small telescope and a good deal of leisure, began for his own amusement to note their number every day. He commenced in 1826, and with German patience observed daily for forty years. He first found that the spots grew more numerous in 1830, when there was no single day without one; then the number declined very rapidly, till in 1833 they were about gone; then they increased in number again till 1838, then again declined; and so on, till it became evident that sun-spots do not come and go by chance, but run through a cycle of growth and disappearance, on the average about once in every eleven years. While amusing himself with his telescope, an important sequence

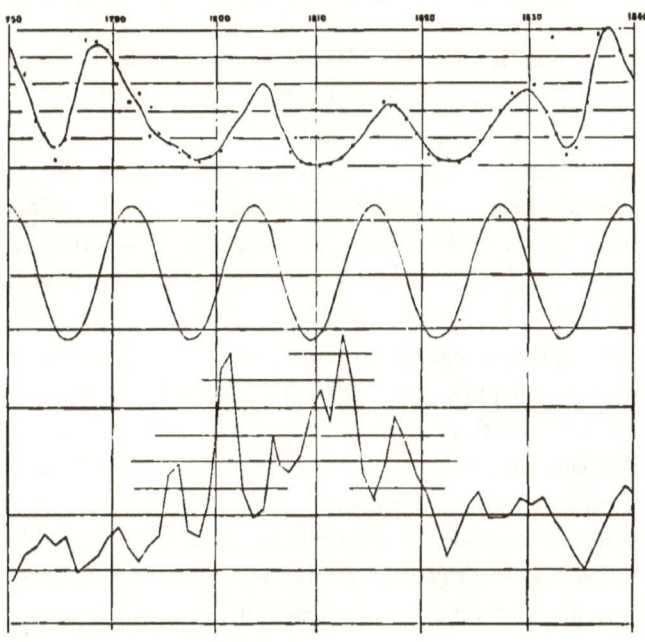

FIG. 48. — SUN-SPOTS AND PRICE OF GRAIN. (FROM "OBSERVATIONS OF SOLAR SPOTS.")

in Nature had thus been added to our knowledge by the obscure Hofrath Schwabe, who indeed compares himself to Saul, going out to seek his father's asses and finding a kingdom. Old records made before Schwabe's time have since been hunted up, so that we have a fairly connected history of the sun's surface for nearly a hundred and fifty years; and the years when spots will be plentiful or rare can now be often predicted from seeing what has been in the past. Thus I may venture to say that the spots, so frequent in 1885, will have probably nearly disappeared in 1888, and will be probably very plentiful in 1894. I do not know at all why this is likely to happen; I only know that it has repeatedly happened at corresponding periods in the past.

"Now," it may be asked, "have these things any connection with weather changes, and is it of any practical advantage to know if they have?"

Would it be, it may be answered, of any practical interest to a merchant in bread-stuffs to have private information of a reliable character that crops the world over would be fine in 1888 and fail in 1894? The exclusive possession of such knowledge might plainly bring "wealth beyond the dreams of avarice" to the user; or, to ascend from the lower ground of personal interest to the higher aims of philanthropy and science, could we predict the harvests, we should be armed with a knowledge that might provide against coming years of famine, and make life distinctly happier and easier to hundreds of millions of toilers on the earth's surface.

"But can we predict?" We certainly cannot till we have, at any rate, first shown that there is a connection between sun-spots and the weather. Since we know nothing of the ultimate causes involved, we can only at present, as I say, collect records of the changes there, and compare them with others of the changes here, to see if there is any significant coincidence. To avoid columns of figures, and yet to enable the reader to judge for himself in some degree of the evidence, I will give the results of some of these records represented graphically by curves, like those which he may perhaps remember to have seen used to show the fluctuations in the value of gold and grain, or of stocks in the stock-market. It is only fair to say that mathematicians used this method long before it was ever heard of by business men, and that the stockbrokers borrowed it from the astronomers, and not the astronomers from them.

In Fig. 48, from Carrington's work, each horizontal space represents ten years of time, and the figures in the upper part represent the fluctuations of the sun-spot curve. In the middle curve, variations in vertical distances correspond to differences in the distance from the sun of the planet Jupiter, the possibility of whose influence on sun-spot periods can thus be examined. In the third and lowest, suggested by Sir William Herschel, the figures at the side are proportional to the price of wheat in the English market, rising when wheat ruled high, falling when it was cheap. In all three curves one-tenth of a horizontal spacing along the top or bottom corresponds to one year; and in this way we have at a glance the condensed result of observations and statistics for sixty years, which otherwise stated would fill volumes. The result is instructive in more ways than one. The variations of Jupiter's distance certainly do present a striking coincidence with the changes in spot frequency, and this may indicate a real connection between the phenomena; but before we decide that it does so, we must remember that the number of cycles of change presented by the possible combination of planetary periods is all but infinite. Thus we might safely undertake, with study enough, to find a curve, depending solely on certain planetary configurations, which yet would represent with quite striking agreement for a time the rise and fall in any given railroad stock, the relative numbers of Democratic and Republican congressmen from year to year, or anything else with which the heavenly bodies have in reality as little to do. The third curve (meant by the price of wheat to test the possible influence of sun-spots on years of good or bad harvests) is not open to the last objection, but involves a fallacy of another kind. In fact the price of wheat depends on many things quite apart from the operations of Nature, — on wars and legislation, for instance; and here the great rise in the first years of the century is as clearly connected with the great continental wars of the first Napoleon, which shut up foreign ports, as the sudden fall about 1815, the year of Waterloo, is with the subsequent peace. Meanwhile an immense amount of labor has been spent in making tables of the weather, and of almost every conceivable earthly phenomenon which may be supposed to have a similar periodic character, with very doubtful success, nearly every one having brought out some result which might be plausible if it stood alone, but which is apt to be contradicted by the others. For instance, Mr. Stone, at the Cape of Good Hope, and Dr. Gould, in South America, consider that the observations taken at those places show a little diminution of the earth's temperature (amounting to one or two degrees) at a sun-spot maximum. Mr. Chambers concludes, from twenty-eight years' observations, that the hottest are those of most sun-spots. So each of these contradicts the other. Then we have Gelinck, who, from a study of numerous observations, concludes that all are wrong together, and that there is really no change in either way.

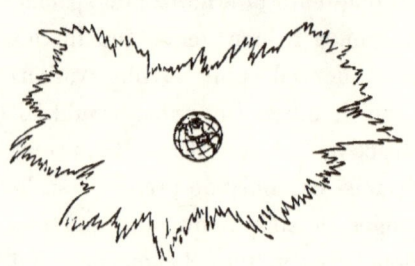

FIG. 40. — SUN-SPOT OF NOV. 16, 1882, AND EARTH.

I might go on citing names with no better result. One observer tabulates observations of terrestrial temperature, or rain-fall, or barometer, or ozone; another, the visitations of Asiatic cholera; while still another (the late Professor Jevons) tabulates commercial crises with the serious attempt to find a connection between the sun-spots and business panics. Of making such cycles there is no end, and much study of them would be a weariness I will not inflict.

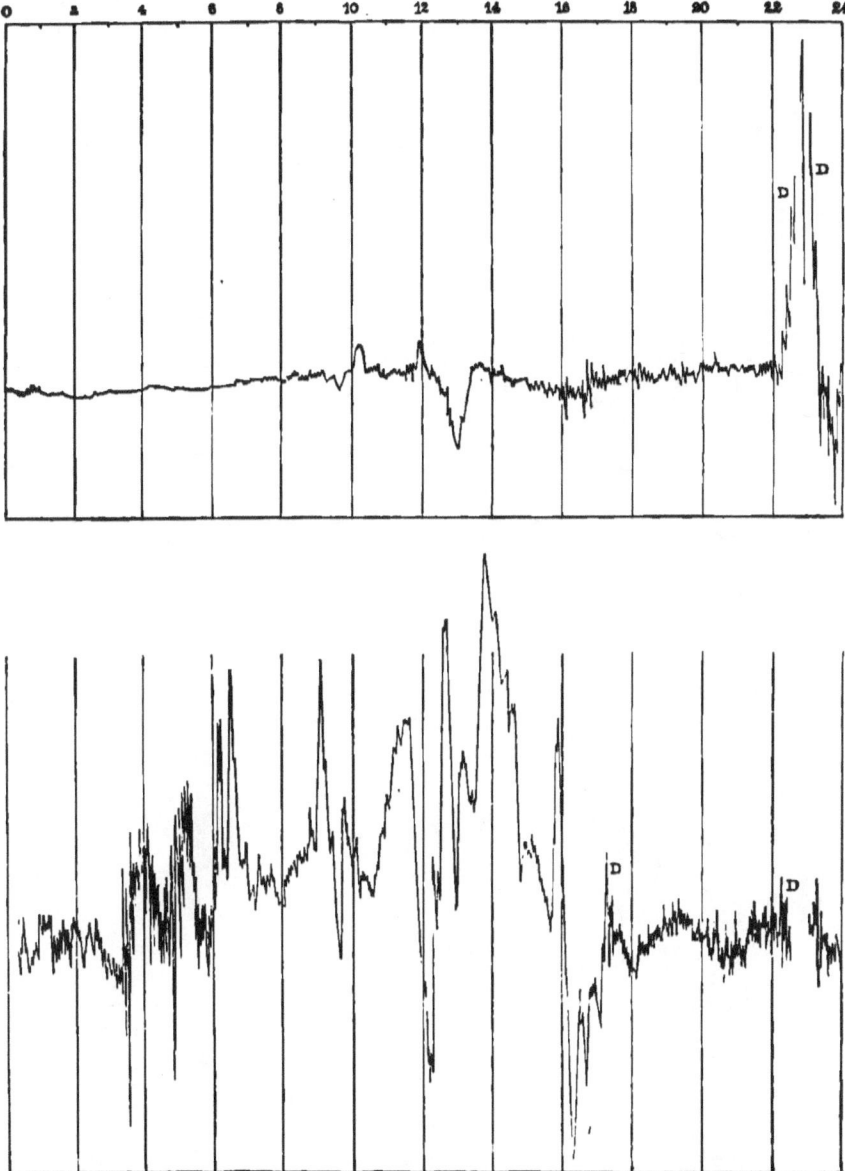

FIG. 50.— GREENWICH RECORD OF DISTURBANCE OF MAGNETIC NEEDLE, NOV. 16 AND 17, 1882.

Our own conclusion is, that from such investigations of terrestrial changes nothing is yet certainly known with regard to the influence of sun-spots on the weather. There is, however, quite another way; that is, to measure their effect at the origin in the sun itself. The sun-spot is cooler than the rest of the surface, and it might be thought that when there are many the sun would give less heat. As far as the spots themselves are concerned, this is so, but in a very small degree. I have been able to ascertain how much this deprivation of heat amounts to, and find it is a real but a most insignificant quantity, rising to about two-thirds of one degree Fahrenheit every eleven years. This, it will be remembered, is the direct effect of the spots considered merely as so many cool patches on the surface, and it does not imply that when there are most spots the sun will necessarily give less heat. In fact there may be a compensating action accompanying them which makes the radiation greater than when they are absent. I will not enter on a detailed explanation, but only say that in the best judgment I can form by a good deal of study and direct experiment, there is no certain evidence that the sun is hotter at one time than at another.

If we investigate, however, the connection between spots and terrestrial magnetic disturbances, we shall find altogether more satisfactory testimony. This evidence is of all degrees of strength, from probability up to what may be called certainty, and it is always obtained, not by *a priori* reasoning, but by the comparison of independent observations of something which has happened on the sun and on the earth. We will first take an instance of what we consider the weakest degree of evidence (weak, that is, when any such single case is considered), and we do so by simply quoting textually three records which were made at nearly the same time in different parts of the world in 1882.

A certain spot had been visible on the sun at intervals for some weeks; but when on the 16th of November a glimpse was caught of it after previous days of cloudy weather, the observer, it will be seen, is struck by the great activity going on in it, and, though familiar with such sights, describes this one as "magnificent."

1. From the daily record at the Allegheny Observatory, November 16, 1882: —

"Very large spot on the sun; . . . great variety of forms; inrush from S. E. to S. W.; tendency to cyclonic action at several points. The spot is apparently near its period of greatest activity. A magnificent sight."

At the same time a sketch was commenced which was interrupted by the cloudy weather of this and following days. The outline of the main spot only is here given (Fig. 49). Its area, as measured at Allegheny, was 2,200,000,000 square miles; at Greenwich its area, inclusive of some outlying portions, was estimated on the same day to be 2,600,000,000 square miles. The earth is shown of its relative size upon it, to give a proper idea of the scale.

2. From the "New York Tribune" of November 18th (describing what took place in the night preceding the 17th): —

AN ELECTRIC STORM.
TELEGRAPH WIRES GREATLY AFFECTED.
THE DISTURBANCE WIDE-SPREAD.

. . . At the Mutual Union office the manager said, "Our wires are all running, but very slowly. There is often an intermission of from one to five minutes between the words of a sentence. The electric storm is general as far as our wires are concerned." . . . The cable messages were also delayed, in some cases as much as an hour.

The telephone service was practically useless during the day.

WASHINGTON, *Nov.* 17. — A magnetic storm of more than usual intensity began here at an early hour this morning, and has continued with occasional interruptions during the day, seriously interfering with telegraphic communication. . . . As an experiment one of the wires of the Western Union Telegraph Company was worked between Washington and Baltimore this afternoon with the terrestrial current alone, the batteries having been entirely detached.

CHICAGO, *Nov.* 17.— An electric storm of the greatest violence raged in all the territory to points beyond Omaha. . . . The switch-board here has been on fire a dozen times during the forenoon. At noon only a single wire out of fifteen between this city and New York was in operation.

And so on through a column.

3. In Fig. 50 we give a portion of the automatic trace of the magnetic needles at Greenwich.[1] These needles are mounted on massive piers in the cellars of the observatory, far removed from every visible source of disturbance, and each carries a small mirror, whence a spot of light is reflected upon a strip of photographic paper, kept continually rolling before it by clockwork. If the needle is still, the moving strip of paper will have a straight line on it, traced by the point of light, which is in this case motionless. If the needle swings to the right or left, the light-spot vibrates with it, and the line it traces becomes sinuous, or more and more sharply zigzagged as the needle shivers under the unknown forces which control it.

The upper part of Fig. 50 gives a little portion of this automatic trace on November 16th before the disturbance began, to show the ordinary daily record, which should be compared with the violent perturbation occurring simultaneously with the telegraphic disturbance in the United States. We may, for the reader's convenience, remark that as the astronomical day begins twelve hours later than the civil day, the approximate Washington mean times, corresponding to the Greenwich hours after

---

[1] It appears here through the kindness of the Astronomer Royal. We regret to say that American observers are dependent on the courtesy of foreign ones in such matters, the United States having no observatory where such records of sun-spots and magnetic variation are systematically kept.

twelve, are found by adding one to the days and subtracting seventeen from the hours. Thus "November 16th, twenty-two hours" corresponds in the eastern United States nearly to five o'clock in the morning of November 17th.

The Allegheny observer, it will be remembered, in his glimpse of the spot on November 16th, was struck with the great activity of the internal motions then going on in it. The Astronomer Royal states that a portion of the spot became detached on November 17th or 18th, and that several small spots which broke out in the immediate neighborhood were seen for the first time on the photographs taken November 17th, twenty-two hours.

"Are we to conclude from this," it may be asked, "that what went on in the sun was the cause of the trouble on the telegraph wires?" I think we are not at all entitled to conclude so from this instance *alone;* but though in one such case, taken by itself, there is nothing conclusive, yet when such a degree of coincidence occurs again and again, the habitual observer of solar phenomena learns to look with some confidence for evidence of electrical disturbance here following certain kinds of disturbance there, and the weight of this part of the evidence is not to be sought so much in the strength of a single case, as in the multitude of such coincidences.

We have, however, not only the means of comparing sunspot *years* with years of terrestrial electric disturbance, but individual instances, particular *minutes* of sun-spot changes, with particular minutes of terrestrial change; and both comparisons are of the most convincing character.

First, let us observe that the compass needle, in its regular and ordinary behavior, does not point constantly in any one direction through the day, but moves a very little one way in the morning, and back in the afternoon. This same movement, which can be noticed even in a good surveyor's compass, is called the "diurnal oscillation," and has long been known. It has been known, too, that its amount altered from one year to another; but since Schwabe's observations it has been found that the changes in this variation and in the number of the spots went on together. The coincidences which we failed to note in the comparison of the spots with the prices of grain are here made out with convincing clearness, as the reader will see by a simple inspection of this chart (Fig. 51, taken from Professor Young's

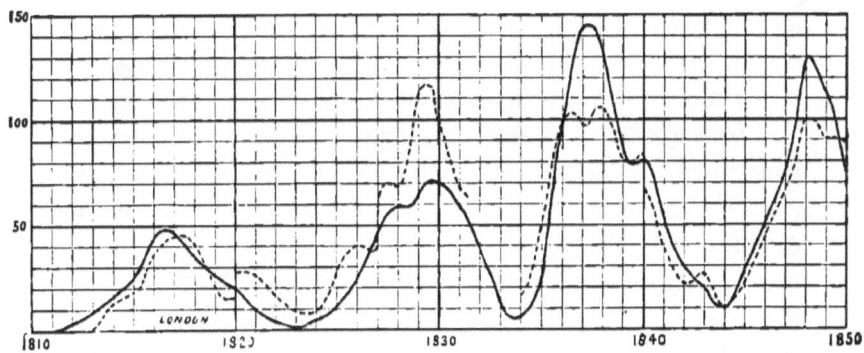

FIG. 51.—SUN-SPOTS AND MAGNETIC VARIATIONS.

work), where the horizontal divisions still denote years, and the height of the continuous curve the relative number of spots, while the height of the dotted curve is the amount of the magnetic variation. Though we have given but a part of the curve, the presumption from the agreement in the forty years alone would be a strong one that the two effects, apparently so widely remote in their nature, are really due to a common cause.

Here we have compared years with years; let us next compare minutes with minutes. Thus, to cite (from Mr. Proctor's work) a well-known instance: On Sept. 1, 1869, at eighteen minutes past eleven, Mr. Carrington, an experienced solar observer, suddenly saw in the sun something brighter than the sun, — two patches of light, breaking out so instantly and so intensely that his first thought was that daylight was entering through a hole in the darkening screen he used. It was immediately, however, made certain that something unusual was occurring in the sun itself, across which the brilliant spots were moving, travelling thirty-five thousand miles in five minutes, at the end of which time (at twenty-three minutes past eleven) they disappeared from sight. By good fortune, another observer a few miles distant saw and independently described the same phenomenon; and as the minute had been noted, it was immediately afterward found that recording instruments registered a magnetic disturbance at the same time, — "at the very moment," says Dr. Stewart, the director of the observatory at Kew.

"By degrees," says Sir John Herschel, "accounts began to pour in of . . . great electro-magnetic disturbances in every part of the world. . . . At Washington and Philadelphia, in America, the telegraphic signal men received severe electric shocks. At Boston, in North America, a flame of fire followed the pen of Bain's electric telegraph." (Such electric disturbances, it may be mentioned, are called "electric storms," though when they occur the weather may be perfectly serene to the eye. They are shown also by rapid vibrations of the magnetic needle, like those we have illustrated.)

On Aug. 3, 1872, Professor Young, who was observing at Sherman in the Rocky Mountains, saw three notable paroxysms in the sun's chromosphere, jets of luminous matter of intense brilliance being projected at 8h. 45m., 10h. 30m., and 11h. 50m. of the local time. "At dinner," he says, "the photographer of the party, who was making our magnetic observations, told me, before knowing anything about what I had been observing, that he had been obliged to give up work, his magnet having swung clear off the limb." Similar phenomena were observed August 5th. Professor Young wrote to England, and received from Greenwich and Stonyhurst copies of the automatic record, which he gives, and which we give in Fig. 52. After allowing for difference of longitude, the reader who will take the pains to compare them may see for himself that both show a jump of the needles in the cellars at Greenwich at the same *minute* in each of the four cases of outburst in the Rocky Mountains.

While we admit that the evidence in any single case is rarely so conclusive as in these; while we agree that the spot is not so

FIG. 52.— GREENWICH MAGNETIC OBSERVATIONS, AUG. 3 AND 5, 1872.

much the cause of the change as the index of some other solar action which does cause it; and while we fully concede our present ignorance of the nature of this cause, — we cannot refuse to accept the cumulative evidence, of which a little has been submitted.

It is only in rare cases that we can feel quite sure; and yet, in regard even to one of the more common and less conclusive ones, we may at least feel warranted in saying that if the reader forfeited a business engagement or missed an invitation to dinner through the failure of the telegraph or telephone on such an occasion as that of the 17th of November, 1882, the far-off sun-spot was not improbably connected with the cause.

Probably we should all like to hear some at least equally positive conclusion about the weather also, and to learn that there was a likelihood of our being able to predict it for the next year, as the Signal Service now does for the next day; but there is at present no such likelihood. The study of the possible connection between sun-spots and the weather is, nevertheless, one that will always have great interest to many; for even if we set its scientific aim aside and consider it in its purely utilitarian aspect, it is evident that the knowledge how to predict whether coming harvests would be good or bad, would enable us to do for the whole world what Joseph's prophetic vision of the seven good and seven barren years did for the land of Egypt, and confer a greater power on its discoverer than any sovereign now possesses. There is something to be said, then, for the cyclists; for if their zeal does sometimes outrun knowledge, their object is a worthy one, and their aims such as we can sympathize with, and of which none of us can say that there is any inherent impossibility in them, or that they may not conceivably yet lead to something. Let us not, then, treat the inquirer who tries to connect panics on 'Change with sun-spots as a mere lunatic; for there is this amount of reason in his theory, that the panics, together with the general state of business, are connected in some obscure way with the good or bad harvests, and these again in some still obscurer way with changes in our sun.

We may leave, then, this vision of forecasting the harvests and the markets of the world from a study of the sun, as one of the fair dreams for the future of our science. Perhaps the dream will one day be realized. Who knows?

IV.

THE SUN'S ENERGY (*Continued*).

IF we paused on the words with which our last chapter closed, the reader might perhaps so far gather an impression that the whole all-important subject of the solar energy was involved in mystery and doubt. But if it be indeed a mystery when considered in its essence, so are all things; while regarded separately in any one of its terrestrial effects of magnetic or chemical action, or of light or heat, it may seem less so. Since there is not room to consider all these aspects, let us choose the last, and look at this energy in its familiar form of the *heat* by which we live.

We, the human race, are warming ourselves at this great fire which called our bodies into being, and when it goes out we shall go too. What is it? How long has it been? How long will it last? How shall we use it?

To look across the space of over ninety million miles, and to try to learn from that distance the nature of the solar heat, and how it is kept up, seemed to the astronomers of the last century a hopeless task. The difficulty was avoided rather than met by the doctrine that the sun was pure fire, and shone because "it was its nature to." In the Middle Ages such an idea was universal; and along with it, and as a logical sequence of it, the belief was long prevalent that it was possible 'to make another such flame here, in the form of a lamp which should burn forever and radiate light endlessly without exhaustion. With the philosopher's stone, which was to transmute lead into gold, this perpetual lamp formed a prime object of research for the alchemist and student of magic.

We recall the use which Scott has made of the belief in this product of "gramarye" in the "Lay of the Last Minstrel," where it is sought to open the grave of the great wizard in Melrose Abbey. It is midnight when the stone which covers it is heaved away, and Michael's undying lamp, buried with him long ago, shines out from the open tomb and illuminates the darkness of the chancel.

> "I would you had been there to see
> The light break forth so gloriously;
> That lamp shall burn unquenchably
> Until the eternal doom shall be,"

says the poet. Now we are at liberty to enjoy the fiction as a fiction; but if we admit that the art which could make such a lamp would indeed be a black art, which did not work under Nature's laws, but against them, then we ought to see that as the whole conception is derived from the early notion of a miraculous constitution of the sun, the idea of an eternal self-sustained sun is no more permitted to us than that of an eternal self-sustained lamp. We must look for the cause of the sun's heat in Nature's laws, and we know those laws chiefly by what we see here.

Before examining the source of the sun's heat, let us look a little more into its amount. To find the exact amount of heat which it sends out is a very difficult problem, especially if we are to use all the refinements of the latest methods in determining it. The underlying principle, however, is embodied in an old method, which gives, it is true, rather crude results, but by so simple a treatment that the reader can follow it readily, especially if unembarrassed with details, in which most of the actual trouble lies. We must warn him in advance that he is going to be confronted with a kind of enormous sum in multiplication, for whose general accuracy he may, however, trust to us if he pleases. We have not attempted exact accuracy, because it is more convenient for him that we should deal with round numbers.

The apparatus which we shall need for the attack of this great problem is surprisingly simple, and moderate in size. Let us begin by finding how much sun-heat falls in a small known area. To do this we take a flat, shallow vessel, which is to be filled with water. The amount it contains is usually a hundred cubic centimetres (a centimetre being nearly four-tenths of an inch), so that if we imagine a tiny cubical box about as large as a backgammon die, or, more exactly, having each side just the size of this (Fig. 53), to be filled and emptied into the vessel one hundred times, we shall have a precise idea of its limited capacity.

FIG. 53.—ONE CUBIC CENTIMETRE.

Into this vessel we dip a thermometer, so as to read the temperature of the water, seal all up so that the water shall not run out, and expose it so that the heat at noon falls perpendicularly on it. The apparatus is shown in Fig. 54, attached to a tree. The stem of the instrument holds the thermometer, which is upside down, its bulb being in the water-vessel. Now, all the sun's rays do not reach this vessel, for some are absorbed by our atmosphere; and all the heat which falls on it does not stay there, as the water loses part of it by the contact of the air with the box outside, and in other ways. When allowance is made for these losses, we find that the sun's heat, if all retained, would have raised the temperature of the few drops of water which would fill a box the size of our little cube (according to these latest observa-

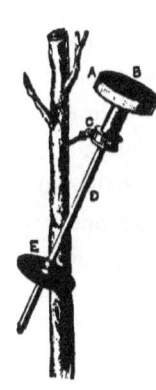

FIG. 54.—POUILLET'S PYRHELIOMETER.

tions) nearly three degrees of the centigrade thermometer in one minute, — a most insignificant result, apparently, as a measure of what we have been told is the almost infinite heat of the sun! But if we think so, we are forgetting the power of numbers, of which we are about to have an illustration as striking in its way as that which Archimedes once gave with the grains of sand.

There is a treatise of his extant, in which he remarks (I cite from memory) that as some people believe it possible for numbers to express a quantity as great as that of the grains of sand upon the sea-shore, while others deny this, he will show that they can express one even larger. To prove this beyond dispute, he begins by taking a small seed, beside which he ranges single grains of sand in a line, till he can give the number of these latter which equal its length. Next he ranges seeds beside each other till their number makes up the length of a span; then he counts the spans in a stadium, and the stadia in the whole world as known to the ancients, at each step expressing his results in a number certainly *greater* than the number of sand-grains which the seed, or the span, or the stadium, or finally the whole world, is thus successively shown to contain. He has then already got a number before his reader's eyes demonstrably larger than that of all the grains of sand on the sea-shore; yet he does not stop, but steps off the earth into space, to calculate and express a number *greater* than that of all the grains of sand which would fill a sphere embracing the earth and the sun!

We are going to use our little unit of heat in the same way, for (to calculate in round figures and in English measure) we find that we can set over nine hundred of these small cubes side by side in a square foot, and, as there are 28,000,000 feet in a square mile, that the latter would contain 25,000,000,000 of the cubes, placed side by side, touching each other, like a mosaic pavement. We find also, by weighing our little cup, that we should need to fill and empty it almost exactly a million times to exhaust a tank containing a ton of water. The sun-heat falling on one square mile corresponds, then, to over seven hundred and fifty tons of water raised *every minute* from the freezing-point to boiling, which already is becoming a respectable amount!

But there are 49,000,000 square miles in the cross-section of the earth exposed to the sun's rays, which it would therefore need 1,225,000,000,000,000,000 of our little dies to cover one deep; and therefore in each *minute* the sun's heat falling on the earth would raise to boiling 37,000,000,000 tons of water.

We may express this in other ways, as by the quantity of ice it would melt; and as the heat required to melt a given weight of ice is $\frac{10}{100}$ of that required to bring as much water from the freezing to the boiling point, and as the whole surface of the earth, including the night side, is four times the cross-section exposed to the sun, we find, by taking 526,000 minutes to a year, that the sun's rays would melt in the year a coating of ice over the whole earth more than one hundred and sixty feet thick.

We have ascended already from our small starting-point to numbers which express the heat that falls upon the whole planet, and enable us to deal, if we wish, with questions relating to the glacial epochs and other changes in its history. We have done this by referring at each step to the little cube which we have carried along with us, and which is the foundation of all the rest; and we now see why such exactness in the first determination is needed, since any error is multiplied by enormous numbers. But now we too are going to step off the earth and to deal with numbers which we can still express in the same way if we choose, but which grow so large thus stated that we will seek some greater term of comparison for them. We have just seen the almost incomprehensible amount of heat which the sun must send the earth in order to warm its oceans and make green its continents; but how little this is to what passes us by! The earth as it moves on in its annual path continually comes into new regions, where it finds the same amount of heat already pouring forth; and this same amount still continues to fall into the empty space we have just quitted, where there is no one left to note it, and where it goes on in what seems to us utter waste. If, then, the whole annual orbit were set close with globes like ours, and strung with worlds like beads upon a ring, each would receive the same enormous amount the earth does now. But this is not all; for not only along the orbit, but above and below it, the sun sends its heat in seemingly incredible wastefulness, the final amount being expressible in the number of *worlds* like ours that it could warm like ours, which is 2,200,000,000.

We have possibly given a surfeit of such numbers, but we cannot escape or altogether avoid them when dealing with this stupendous outflow of the solar heat. They are too great, perhaps, to convey a clear idea to the mind, but let us before leaving them try to give an illustration of their significance.

Let us suppose that we could sweep up from the earth all the ice and snow on its surface, and, gathering in the accumulations which lie on its Arctic and Antarctic poles, commence building with it a tower greater than that of Babel, fifteen miles in diameter, and so high as to exhaust our store. Imagine that it could be preserved untouched by the sun's rays, while we built on with the accumulations of successive winters, until it stretched out 240,000 miles into space, and formed an ice-bridge to the moon, and that then we concentrated on it the sun's whole radiation, neither more nor less than that which goes on every moment. In *one* second the whole would be gone, melted, boiled, and dissipated in vapor. And this is the rate at which the solar heat is being (to human apprehension) *wasted!*

Nature, we are told, always accomplishes her purpose with the least possible expenditure of energy. Is her purpose here, then, something quite independent of man's comfort and happiness? Of the whole solar heat, we have just seen that less than $\frac{1}{2,000,000}$, — less, that is, than the one twenty-thousandth part of one per cent, — is made useful to us. "But may there not be other planets on which intelligent life exists, and where this heat, which passes us by, serves other beings than ourselves?" There *may* be; but if we could suppose all the other planets of the solar system to be inhabited, it would help the matter very little; for the whole together intercept so little of the great sum, that all of it which Nature bestows on man is still as nothing to what she bestows on some end — if end there be — which is to us as yet inscrutable.

How is this heat maintained? Not by the miracle of a perpetual self-sustained flame, we may be sure. But, then, by what fuel is such a fire fed? There can be no question of simple burning, like that of coal in the grate, for there is no source of supply adequate to the demand. The State of Pennsylvania, for instance, is underlaid by one of the richest coal-fields of the world, capable of supplying the consumption of the whole country at its present rate for more than a thousand years to come. If the source of the solar heat (whatever that is) were withdrawn, and we were enabled to carry this coal there, and shoot it into the solar furnace fast enough to keep up the known heat-supply, so that the solar radiation would go on at just its actual rate, the time which this coal would last is easily calculable. It would not last days or hours, but the whole of these coal-beds would demonstrably be used up in rather less than one one-thousandth of a second! We find by a similar calculation that if the sun were itself one solid block of coal, it would have burned out to the last cinder in less time than man has certainly been on the earth. But during historic times there has as surely been no noticeable diminution of the sun's heat, for the olive and the vine grow just as they did three thousand years ago, and the hypothesis of an actual burning becomes untenable. It has been supposed by some that meteors striking the solar surface might generate heat by their impact, just as a cannon-ball fired against an armor-plate causes a flash of light, and a heat so sudden and intense as to partly melt the ball at the instant of concussion. This is probably a real source of heat-supply so far as it goes, but it cannot go very far; and, indeed, if our whole world should fall upon the solar surface like an immense projectile, gathering speed as it fell, and finally striking (as it would) with the force due to a rate of over three hundred miles a second, the heat developed would supply the sun for but little more than sixty years.[1]

It is not necessary, however, that a body should be moving rapidly to develop heat, for arrested motion always generates it, whether the motion be fast or slow, though in the latter case the mass arrested must be larger to produce the same result. It is in the slow settlement of the sun's own substance toward its centre, as it contracts in cooling, that we find a sufficient cause for the heat developed.

This explanation is often unsatisfactory to those who have not studied the subject, because the fact that heat is so generated is not made familiar to most of us by observation.

Perhaps the following illustration will make the matter plainer. When we are carried up in a lift, or elevator, we know well enough that heat has been expended under the boiler of some engine to drag us up against the power of gravity. When the elevator is at the top of its course, it is ready to give out in descending just the same amount of power needed to raise it, as we see by its drawing up a nearly equal

---

[1] These estimates differ somewhat from those of Helmholtz and Tyndall, as they rest on later measures.

counterpoise in the descent. It can and must give out in coming down the power that was spent in raising it up; and though there is no practical occasion to do so, a large part of this power could, if we wished, be actually recovered in the form of heat again. In the case of a larger body, such as the pyramid of Ghizeh, which weighs between six and seven million tons, all the furnaces in the world, burning coal under all its engines, would have to supply their heat for a measurable time to lift it a mile high; and then, if it were allowed to come down, whether it fell at once or were made to descend with imperceptible slowness, by the time it touched the earth the same heat would be given out again.

Perhaps the fact that the sun is gaseous rather than solid makes it less easy to realize the enormous weight which is consistent with this vaporous constitution. A cubic mile of hydrogen gas (the lightest substance known) would weigh much more at the sun's surface than the Great Pyramid does here, and the number of these cubic miles in a stratum one mile deep below its surface is over 2,000,000,000,000! This alone is enough to show that as they settle downward as the solar globe shrinks, here is a *possible* source of supply for all the heat the sun sends out. More exact calculation shows that it *is* sufficient, and that a contraction of three hundred feet a year (which in ten thousand years would make a shrinkage hardly visible in the most powerful telescope) would give all the immense outflow of heat which we see.

There is an ultimate limit, however, to the sun's shrinking, and there must have been some bounds to the heat he can already have thus acquired; for — though the greater the original diameter of his sphere, the greater the gain of heat by shrinking to its present size — if the original diameter be supposed as great as possible, there is still a finite limit to the heat gained.

Suppose, in other words, the sun itself and all the planets ground to powder, and distributed on the surface of a sphere whose radius is infinite, and that this matter (the same in amount as that constituting the present solar system) is allowed to fall together at the centre. The actual shrinkage cannot possibly be greater than in this extreme case; but even in this practically impossible instance, it is easy to calculate that the heat given out would not support the *present* radiation over eighteen million years, and thus we are enabled to look back over past time, and fix an approximate limit to the age of the sun and earth.

We say "present" rate of radiation, because, so long as the sun is purely gaseous, its temperature rises as it contracts, and the heat is spent faster; so that in early ages before this temperature was as high as it is now, the heat was spent more slowly, and what could have lasted "only" eighteen million years at the present rate might have actually spread over an indefinitely greater time in the past; possibly covering more than all the æons geologists ask for.

If we would look into the future, also, we find that at the present rate we may say that the sun's heat-supply is enough to last for some such term as four or five million years before it sensibly fails. It is certainly remarkable that by the aid of our science man can look out from this "bank and shoal of time," where his fleeting existence is spent, not only back on the almost infinite lapse of ages past, but that he can forecast with some sort of assurance what is to happen in an almost infinitely distant future, long after the human race itself will have disappeared from its present home. But so it is, and we may say — with something like awe at the meaning to which science points — that the whole future radiation cannot last so long as ten million years. Its probable life in its present condition is covered by about thirty million years. No reasonable allowance for the fall of meteors or for all other known causes of supply could possibly at the present rate of radiation raise the whole term of its existence to sixty million years.

This is substantially Professor Young's view, and he adds: —

"At the same time it is, of course, impossible to assert that there has been no catastrophe in the past, no collision with some wandering star . . . producing a shock which might in a few hours, or moments even, restore the wasted energy of ages. Neither is it wholly safe to assume that there may not be ways, of which we as yet have no conception, by which the energy apparently lost in space may be returned. But the whole course and tendency of Nature, so far as science now makes out, points backward to a beginning and forward to an end. The present order of things seems to be bounded both in the past and in the future by terminal catastrophes which are veiled in clouds as yet impenetrable."

There is another matter of interest to us as dwellers on this planet, connected not with the amount of the sun's heat so much as with the degree of its temperature; for it is almost certain that a very little fall in the temperature will cause an immense and wholly disproportionate diminution of the heat-supply. The same principle may be observed in more familiar things. We can, for instance, warm quite a large house by a very small furnace, if we urge this (by a wasteful use of coal) to a dazzling white heat. If we now let the furnace cool to half this white-heat temperature, we shall be sure to find that the heat radiated has not diminished in proportion, but out of all proportion, — has sunk, for instance, not only to one-half what it was (as we might think it would do), but to perhaps a twentieth or even less, so that the furnace which heated the house can no longer warm a single room.

The human race, as we have said, is warming itself at the great solar furnace, which we have just seen contains an internal source for generating heat enough for millions of years to come; but we have also learned that if the sun's internal circulation were stopped, the surface would cool and shut up the heat inside, where it would do us no good. The *temperature* of the surface, then, on which the rate of heat-emission depends, concerns us very much; and if we had a thermometer so long that we could dip its bulb into the sun and read the degrees on the stem here, we should find out what observers would very much like to know, and at present are disposed to quarrel about. The difficulty is not in measuring the heat, — for that we have just seen how to do, — but in telling what temperature corresponds to it, since there is no known rule by which to find one from the other. One certain thing is this — that we cannot by any contrivance raise the temperature in the focus of any lens or mirror beyond that of its source (practically we cannot do even so much); we cannot, for instance, by any burning-lens make the image of a candle as hot as the original flame. Whatever a thermometer may read when the candle-heat is concentrated on its bulb by a lens, it would read yet more if the bulb were dipped in the candle-flame itself; and one obvious application of this fact is that though we cannot dip our thermometer in the sun, we know that if we could do so, the temperature would at least be greater than any we get by the largest burning-glass. We need have no fear of making the burning-glass too big; the temperature at its solar focus is *always* and necessarily lower than that of the sun itself.

For some reason no very great burning-lens or mirror has been constructed for a long time, and we have to go back to the eighteenth century to see what can be done in this way. The annexed figure (Fig. 55) is from a wood-cut of the last century, describing the largest burning-lens then or since constructed in France, whose size and mode of use the drawing clearly shows. All the heat falling on the great lens was concentrated on a smaller one, and the smaller one concentrated it in turn, till at the very focus we are assured that iron, gold, and other metals ran like melted butter. In England, the largest burning-lens on record was made about the same time by an optician named Parker for the English Government, who designed it as a present to be taken by Lord Macartney's embassy to the Emperor of China. Parker's lens was three feet in diameter and very massive, being seven inches thick at the

FIG. 55.—BERNIÈRES'S GREAT BURNING-GLASS. (AFTER AN OLD FRENCH PRINT.)

centre. In its focus the most refractory substances were fused, and even the diamond was reduced to vapor, so that the temperature of the sun's surface is at any rate higher than this.

(What became of the French lens shown, it would be interesting to know. If it is still above ground, its fate has been better than that of the English one. It is said that the Emperor of China, when he got his lens, was much alarmed by it, as being possibly sent him by the English with some covert design for his injury. By way of a test, a smith was ordered to strike it with his hammer; but the hammer rebounded from the solid glass, and this was taken to be conclusive evidence of magic in the thing, which was immediately buried, and probably is still reposing under the soil of the Celestial Flowery Kingdom.)

We can confirm the evidence of such burning-lenses as to the sun's high temperature by another class of experiment, which rests on an analogous principle. We can make the comparison between the heat from some artificially heated object and that which would be given out from an equal area of the sun's face. Now, supposing like emissive powers, if the latter be found the hotter, though we cannot tell what its temperature absolutely is, we can at least say that it is greater than that of the thing with which it is compared; so that we choose for comparison the hottest thing we can find, on a scale large enough for the experiment. One observation of my own in this direction I will permit myself to cite in illustration.

Perhaps the highest temperature we can get on a large scale in the arts is that of molten steel in the Bessemer converter. As many may be as ignorant of what this is as I was before I tried the experiment, I will try to describe it.

The "converter" is an enormous iron pot, lined with firebrick, and capable of holding thirty or forty thousand pounds of melted metal; and it is swung on trunnions, so that it can be raised by an engine to a vertical position, or lowered by machinery so as to pour its contents out into a caldron. First the empty converter is inclined, and fifteen thousand pounds of fluid iron streams down into the mouth from an adjacent furnace where it has been melted. Then the engine lifts the converter into an erect position, while an air-blast from a blowing-engine is forced in at the bottom and through the liquid iron, which has combined with it nearly half a ton of

FIG. 56.—A "POUR" FROM THE BESSEMER CONVERTER.

silicon and carbon, — materials which, with the oxygen of the blast, create a heat which leaves that of the already molten iron far behind. After some time the converter is tipped forward, and fifteen hundred pounds more of melted iron is added to that already in it. What the temperature of this last is, may be judged from the fact that though ordinary melted iron is dazzlingly bright, the melted metal in the converter is so much brighter still, that the entering stream is dark brown by comparison, presenting a contrast like that of chocolate poured into a white cup. The contents are now no longer iron, but liquid steel, ready for pouring into the caldron; and, looking from the front down into the inclined vessel, we see the almost blindingly bright interior dripping with the drainage of the metal running down its side, so that the circular mouth, which is twenty-four inches in diameter, presents the effect of a disk of molten metal of that size (were it possible to maintain such a disk in a vertical position). In addition, we have the actual stream of falling metal, which continues nearly a minute, and presents an area of some square feet. The shower of scintillations from this cataract of what seems at first "sunlike" brilliancy, and the area whence such intense heat and light are for a brief time radiated, make the spectacle a most striking one. (See Fig. 56.)

The "pour" is preceded by a shower of sparks, consisting of little particles of molten steel which are projected fully a hundred feet in the direction of the open mouth of the converter. In the line of this my apparatus was stationed in an open window, at a point where its view could be directed down into the converter on one side, and up at the sun on the other. This apparatus consisted of a long photometer-box with a *porte-lumière* at one end. The mirror of this reflected the sun's rays through the box and then on to the pouring metal, tracing their way to it by a beam visible in the dusty air (Fig. 57). In the path of this beam was placed the measuring apparatus, both for heat and light. As the best point of observation was in the line of the blast, a shower of sparks was driven over the instrument and observer at every "pour;" and the rain of wet soot from chimneys without, the bombardment from within, and the moving masses of red-hot iron around, made the experiment an altogether peculiar one.

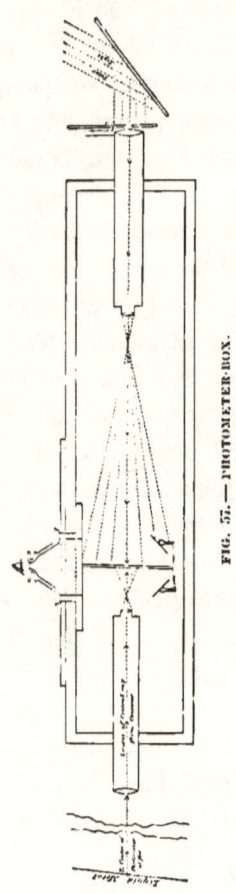

FIG. 57.—PHOTOMETER-BOX.

The apparatus was arranged in such a way that the effect (except for the absorption of its beams on the way) was independent of the size or distance of the sun, and depended on the absolute radiation there, and was equivalent, in fact, to taking a sample piece of the sun's face *of equal size* with the fluid metal, bringing them face to face, and seeing which was the hotter and brighter. The comparison, however, was unfair to the sun, because its rays were in reality partly absorbed by the atmosphere on the way, while those of the furnace were not. Under these circumstances the heat from any single square foot of the sun's surface was found to be *at least* eighty-seven times that from a square foot of the melted metal, while the light from the sun was proved to be, foot for foot, over five thousand times that from the molten steel, though the latter, separately considered, seemed to be itself, as I have said, of quite sunlike brilliancy.

We must not conclude from this that the *temperature* of the sun was five thousand times that of the steel, but we may be certain that it was at any rate a great deal the higher of the two.

FIG. 26.—MOUCHOT'S SOLAR ENGINE. (FROM A FRENCH PRINT.)

It is probable, from all experiments made up to this date, that the solar effective temperature is not less than 3,000 nor more than 30,000 degrees of the centigrade thermometer. Sir William Siemens, whose opinion on any question as to heat is entitled to great respect, thought the lower value nearer the truth, but this is doubtful.

We have, in all that has preceded, been speaking of the sun's constitution and appearance, and have hardly entered on the question of his industrial relations to man. It must be evident, however, that if we derive, as it is asserted we do, almost all our mechanical power from this solar heat, — if our water-wheel is driven by rivers which the sun feeds by the rain he sucks up for them into the clouds, if the coal is stored sun-power, and if, as Stevenson said, it really is the sun who drives our engines, though at second hand, — there is an immense fund of possible mechanical power still coming to us from him which might be economically utilized. Leaving out of sight all our more important relations to him (for, as has been already said, he is in a physical sense our creator, and he keeps us alive from hour to hour), and considering him only as a possible servant to grind our corn and spin our flax, we find that even in this light there are startling possibilities of profit in the study of our subject. From recent measures it appears that from every square yard of the earth exposed perpendicularly to the sun's rays, in the absence of an absorbing atmosphere, there could be derived more than one horse-power, if the heat were all converted into this use, and that even on such a little area as the island of Manhattan, or that occupied by the city of London, the noontide heat is enough, could it all be utilized, to drive all the steam-engines in the world. It will not be surprising, then, to hear that many practical men are turning their attention to this as a source of power, and that, though it has hitherto cost more to utilize the power than it is worth, there is reason to believe that some of the greatest changes which civilization has to bring may yet be due to such investigations. The visitor to the last Paris Exposition may remember an extraordinary machine on the grounds of the Trocadéro, looking like a gigantic inverted umbrella pointed sunward. This was the sun-machine of M. Mouchot, consisting of a great parabolic reflector, which concentrated the heat on a boiler in the focus and drove a steam-engine with it, which was employed in turn to work a printing-press, as our engraving shows (Fig. 58). Because these constructions have been hitherto little more than playthings, we are not to think of them as useless. If toys, they are the toys of the childhood of a science which is destined to grow, and in its maturity to apply this solar energy to the use of all mankind.

Even now they are beginning to pass into the region of practical utility, and in the form of the latest achievement of Mr. Ericsson's ever-young genius are ready for actual work on an economical scale. We present in Fig. 59 his new actually working solar engine, which there is every reason to believe is more efficient than Mouchot's, and probably capable of being used with economical advantage in pumping water in desert regions of our own country. It is pregnant with suggestion of the future, if we consider the growing demand for power in the world, and the fact that its stock of coal, though vast, is strictly limited, in the sense that when it *is* gone we can get absolutely no more. The sun has been making a little every day for millions of years, — so little and for so long, that it is as though time had daily dropped a single penny into the bank to our credit for untold ages, until an enormous fund had been thus slowly accumulated in our favor. We are drawing on this fund like a prodigal who thinks his means endless, but the day will come when our check will no longer be honored, and what shall we do then?

FIG. 59.—ERICSSON'S NEW SOLAR ENGINE, NOW IN PRACTICAL USE IN NEW YORK.

The exhaustion of some of the coal-beds is an affair of the immediate future, by comparison with the vast period of time we have been speaking of. The English coal-beds, it is asserted, will, from present indications, be quite used up in about three hundred years more.

Three hundred years ago, the sun, looking down on the England of our forefathers, saw a fair land of green woods and quiet waters, a land unvexed with noisier machinery than the spinning-wheel, or the needles of the "free maids that weave their threads with bones." Because of the coal which has been dug from its soil, he sees it now soot-blackened, furrowed with railway-cuttings, covered with noisy manufactories, filled with grimy operatives, while the island shakes with the throb of coal-driven engines, and its once quiet waters are churned by the wheels of steamships. Many generations of the lives of men have passed to make the England of Elizabeth into the England of Victoria; but what a moment this time is, compared with the vast lapse of ages during which the coal was being stored! What a moment in the life of the "all-beholding sun," who in a few hundred years — his gift exhausted and the last furnace-fire out — may send his beams through rents in the ivy-grown walls of deserted factories, upon silent engines brown with rust, while the mill-hand has gone to other lands, the rivers are clean again, the harbors show only white sails, and England's "black country" is green once more! To America, too, such a time may come, though at a greatly longer distance.

Does this all seem but the idlest fancy? That something like it will come to pass sooner or later, is a most certain fact — as certain as any process of Nature — if we do not find a new source of power; for of the coal which has supplied us, after a certain time we can get no more.

Future ages may see the seat of empire transferred to regions of the earth now barren and desolated under intense solar heat,

— countries which, for that very cause, will not improbably become the seat of mechanical and thence of political power. Whoever finds the way to make industrially useful the vast sun-power now wasted on the deserts of North Africa or the shores of the Red Sea, will effect a greater change in men's affairs than any conqueror in history has done; for he will once more people those waste places with the life that swarmed there in the best days of Carthage and of old Egypt, but under another civilization, where man no longer shall worship the sun as a god, but shall have learned to make it his servant.

V.

THE PLANETS AND THE MOON.

WHEN we look up at the heavens, we see, if we watch through the night, the host of stars rising in the east and passing above us to sink in the west, always at the same distance and in unchanging order, each seeming a point of light as feeble as the glow-worm's shine in the meadow over which they are rising, each flickering as though the evening wind would blow it out. The infant stretches out its hand to grasp the Pleiades; but when the child has become an old man the "seven stars" are still there unchanged, dim only in his aged sight, and proving themselves the enduring substance, while it is his own life which has gone, as the shine of the glow-worm in the night. They were there just the same a hundred generations ago, before the Pyramids were built; and they will tremble there still, when the Pyramids have been worn down to dust with the blowing of the desert sand against their granite sides. They watched the earth grow fit for man long before man came, and they will doubtless be shining on when our poor human race itself has disappeared from the surface of this planet.

Probably there is no one of us who has not felt this solemn sense of their almost infinite duration as compared with his own little portion of time, and it would be a worthy subject for our thought if we could study them in the light that the New Astronomy sheds for us on their nature. But I must here confine myself to the description of but a few of their number, and speak, not of the infinite multitude and variety of stars, each a self-shining sun, but only of those which move close at hand; for it is not true of quite all that they keep at the same distance and order.

Of the whole celestial army which the naked eye watches, there are five stars which do change their places in the ranks, and these change in an irregular and capricious manner, going about among the others, now forward and now back, as if lost and wandering through the sky. These wanderers were long since known by distinct names, as Mercury, Venus, Mars, Jupiter, and Saturn, and believed to be nearer than the others; and they are, in fact, companions to the earth and fed like it by the warmth of our sun, and like the moon are visible by the sunlight which they reflect to us. With the earliest use of the telescope, it was found that while the other stars remained in it mere points of light as before, these became magnified into disks on which markings were visible, and the markings have been found with our modern instruments, in one case at least, to take the appearance of oceans and snow-capped continents and islands. These, then, are not uninhabitable self-shining suns, but worlds, vivified from the same fount of energy that supplies us, and the possible abode of creatures like ourselves.

"Properly speaking," it is said, "man is the only subject of interest to man;" and if we have cared to study the uninhabitable sun because all that goes on there is found to be so intimately related to us, it is surely a reasonable curiosity which prompts the question so often heard as to the presence of life on these neighbor worlds, where it seems at least not impossible that life should exist. Even the very little we can say in answer to this question will always be interesting; but we must regretfully admit at the outset that it is but little, and that with some planets, like Mercury and Venus, the great telescopes of

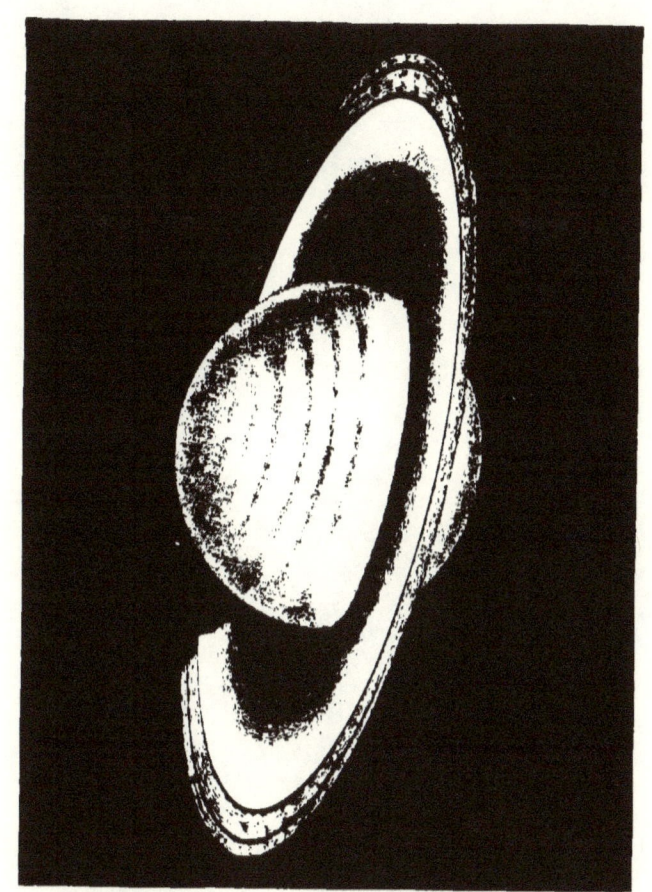

FIG. 59.—SATURN. (FROM A DRAWING BY TROUVELOT).

modern times cannot do much more than those of Galileo, with which our New Astronomy had its beginning.

Let us leave these, then, and pass out to the confines of the planetary system, where we may employ our telescopes to better advantage.

The outer planets, Neptune and Uranus, remain pale disks in the most powerful instruments, the first attended by a single moon, the second by four, barely visible; and there is so very little yet known about their physical features, that we shall do better to give our attention to one of the most interesting objects in the whole heavens, — the planet Saturn, on which we can at any rate see enough to arouse a lively curiosity to know more.

When Galileo first turned his glass on Saturn, he saw, as he thought, that it consisted of three spheres close together, the middle one being the largest. He was not quite sure of the fact, and was in a dilemma between his desire to wait longer for further observation, and his fear that some other observer might announce the discovery if he hesitated. To combine these incompatibilities — to announce it so as to secure the priority, and yet not announce it till he was ready — might seem to present as great a difficulty as the discovery itself; but Galileo solved this, as we may remember, by writing it in the sentence, "Altissimum planetam tergeminum observavi" ("I have observed the highest planet to be triple"), and then throwing it (in the printer's phrase) "into pi," or jumbling the letters, which made the sentence into the monstrous word

SMAJSMRMJLMEBOETALEVMJPVNENVGTTAVJRAS,

and publishing *this*, which contained his discovery, but under lock and key. He had reason to congratulate himself on his prudence, for within two years two of the supposed bodies disappeared, leaving only one. This was in 1612; and for nearly fifty years Saturn continued to all astronomers the enigma which it was to Galileo, till in 1656 it was finally made clear that it was surrounded by a thin flat ring, which when seen fully gave rise to the first appearance in Galileo's small telescope, and when seen edgewise disappeared from its view altogether. Everything in this part of our work depends on the

FIG. 61.—THE EQUATORIAL TELESCOPE AT WASHINGTON.

power of the telescope we employ, and in describing the modern means of observation we pass over two centuries of slow advance, each decade of which has marked some progress in the instrument, to one of its completest types, in the great equatorial at Washington, shown in Fig. 61.

The revolving dome above, the great tube beneath, its massive piers, and all its accessories are only means to carry and direct the great lens at the further end, which acts the part of the lens in our own eye, and forms the image of the thing to be looked at. Galileo's original lens was a single piece of glass, rather smaller than that of our common spectacles; but the lens here is composed of two pieces, each twenty-six inches in diameter, and collects as much light as a human eye would do if over two feet across. But this is useless if the lens is not shaped with such precision as to send every ray to its proper place at the eye-piece, nearly thirty-five feet away; and, in fact, the shape given its surface by the skilful hands of the Messrs. Clark, who made it, is so exquisitely exact that all the light of a star gathered by this great surface is packed at the distant focus into a circle very much smaller than that made by the dot on this *i*, and the same statement may be made of the great Lick glass, which is three feet in diameter, — an accuracy we might call incredible were it not certain. It is with instruments of such accuracy that astronomy now works, and it is with this particular one that some of the observations we are going to describe have been made.

In all the heavens there is no more wonderful object than Saturn, for it preserves to us an apparent type of the plan on which all the worlds were originally made. Let us look at it in this study by Trouvelot (Fig. 60). The planet, we must remember, is a globe nearly seventy thousand miles in diameter, and the outermost ring is over one hundred and fifty thousand miles across, so that the proportionate size of our earth would be over-represented here by a pea laid on the engraving. The belts on the globe show delicate tints of brown and blue, and parts of the ring are, as a whole, brighter than the planet; but this ring, as the reader may see, consists of at least three main divisions, each itself containing separate features. First is the gray outer ring, then the middle one, and next the curious "crape" ring, very much darker than the others, looking like a belt where it crosses the planet, and apparently feebly transparent, for the outline of the globe has been seen (though not very distinctly) *through* it. The whole system of rings is of the most amazing thinness, for it is probably thinner in proportion to its size than the paper on which this is printed is to the width of the page; and when it is turned edgewise to us, it disappears to all but the most powerful telescopes, in which it looks then like the thinnest conceivable line of light, on which the moons have been seen projected, appearing like beads sliding along a golden wire. The globe of the planet casts on the ring a shadow, which is here shown as a broken line, as though the level of the rings were suddenly disturbed. At other times (as in a beautiful drawing made with the same instrument by Professor Holden) the line seems continuous, though curved as though the middle of the ring system were thicker than the edge. The rotation of the ring has been made out by direct observations; and the whole is in motion about the globe, — a motion so smooth and steady that there is no flickering in the shadow "where Saturn's steadfast shade sleeps on its luminous ring."

What is it? No solid could hold together under such conditions; we can hardly admit the possibility of its being a liquid film extended in space; and there are difficulties in admitting it to be gaseous. But if not a solid, a liquid, or a gas, again what can it be? It was suggested nearly two centuries ago that the ring might be composed of innumerable little bodies like meteorites, circling round the globe so close together as to give the appearance we see, much as a swarm of bees at a distance looks like a continuous cloud; and this remains the most plausible solution of what is still in some degree a mystery. Whatever it be, we see in the ring the condition of things which, according to the nebular hypothesis, once pertained to all the planets at a certain stage of their formation; and this,

with the extraordinary lightness of the globe (for the whole planet would float on water), makes us look on it as still in the formative stage of uncondensed matter, where the solid land as yet is not, and the foot could find no resting-place. Astrology figured Saturn as "spiteful and cold,—an old man melancholy;" but if we may indulge such a speculation, modern astronomy rather leads us to think of it as in the infancy of its life, with every process of planetary growth still in its future, and separated by an almost unlimited stretch of years from the time when life under the conditions in which we know it can even begin to exist.

Like this appears also the condition of Jupiter (Fig. 62), the greatest of the planets, whose globe, eighty-eight thousand miles in diameter, turns so rapidly that the centrifugal force causes a visible flattening. The belts which stretch across its disk are of all delicate tints—some pale blue, some of a crimson lake; a sea-green patch has been seen, and at intervals of late years there has been a great oval red spot, which has now nearly gone, and which our engraving does not show. The belts are largely, if not wholly, formed of rolling clouds, drifting and changing under our eyes, though more rarely a feature like the oval spot just mentioned will last for years, an enduring enigma. The most recent observations tend to make us believe that the equatorial regions of Jupiter, like those of the sun, make more turns in a year than the polar ones; while the darkening toward the edge is another sun-like feature, though perhaps due to a distinct cause, and this is beautifully brought out when any one of the four moons which circle the planet passes between us and its face, an occurrence also represented in our figure. The moon, as it steals on the comparatively dark edge, shows us a little circle of an almost lemon-yellow, but the effect of contrast grows less as it approaches the centre. Next (or sometimes before), the disk is invaded by a small and intensely black spot, the shadow of the moon, which slides across the planet's face, the transit lasting long enough for us to see that the whole great globe, serving as a background for the spectacle, has visibly revolved on its axis since we began to gaze. Photography, in the skilful hands of the late Professor Henry Draper, gave us reason to suspect the possibility that a dull light is sent to us from parts of the planet's surface besides what it reflects, as though it were still feebly glowing like a nearly extinguished sun; and, on the whole, a main interest of these features to us lies in the presumption they create that the giant planet is not yet fit to be the abode of life, but is more probably in a condition like that of our earth millions of years since, in a past so remote that geology only infers its existence, and long before our own race began to be. That science, indeed, itself teaches us that such all but infinite periods are needed to prepare a planet for man's abode, that the entire duration of his race upon it is probably brief in comparison.

We pass by the belt of asteroids, and over a distance many times greater than that which separates the earth from the sun, till we approach our own world. Here, close beside it as it were, in comparison with the enormous spaces which intervene between it and Saturn and Jupiter, we find a planet whose size and features are in striking contrast to those of the great globe we have just quitted. It is Mars, which shines so red and looks so large in the sky because it is so near, but whose diameter is only about half that of our earth. This is indeed properly to be called a neighbor world, but the planetary spaces are so immense that this neighbor is at closest still about thirty-four million miles away.

Looking across that great gulf, we see in our engraving (Fig. 63) — where we have three successive views taken at intervals of a few hours — a globe not marked by the belts of Jupiter or Saturn, but with outlines as of continents and islands, which pass in turn before our eyes as it revolves in a little over

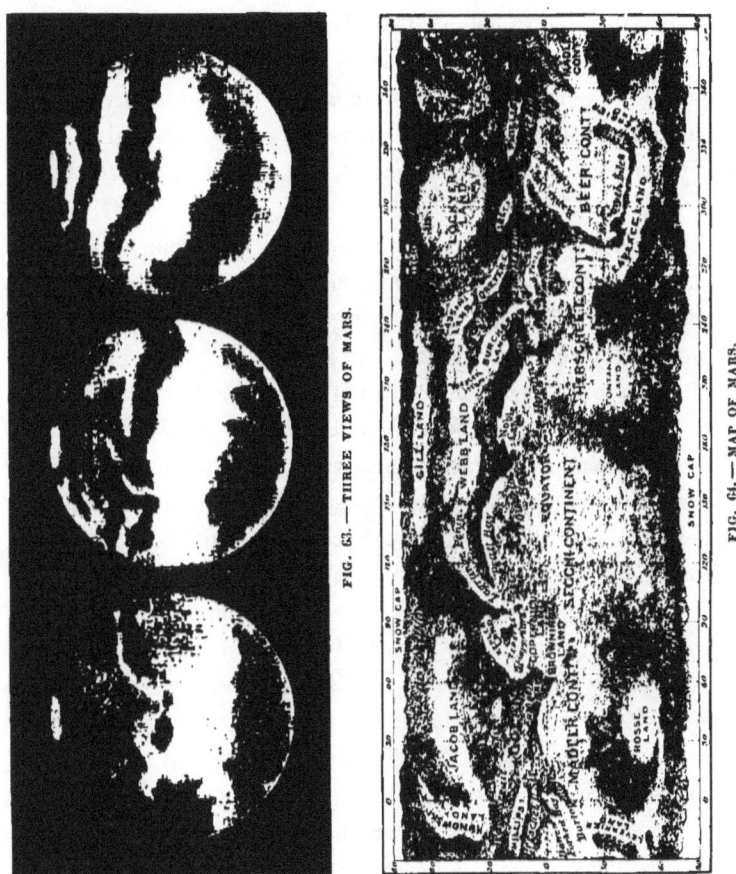

FIG. 63.—THREE VIEWS OF MARS.

FIG. 64.—MAP OF MARS.

twenty-four and a half of our hours, while at either pole is a white spot. Sir William Herschel was the first to notice that this spot increased in size when it was turned away from the sun, and diminished when the solar heat fell on it; so that we have what is almost proof that here is ice (and consequently water) on another world. Then, as we study more, we discern forms which move from day to day on the globe apart from its rotation, and we recognize in them clouds sweeping over the surface, — not a surface of still other clouds below, but of what we have good reason to believe to be land and water.

By the industry of numerous astronomers, seizing every favorable opportunity when Mars comes near, so many of these features have been gathered that we have been enabled to make fairly complete maps of the planet, one of which by Mr. Green is here given (Fig. 64).

Here we see the surface more diversified than that of our earth, while the oceans are long, narrow, canal-like seas, which everywhere invade the land, so that on Mars one could travel almost everywhere by water. These canals have appeared to some observers to exist in pairs, or to resemble two close parallel lines; but this cannot be said to be at present wholly certain. The spectroscope indicates water-vapor in the Martial atmosphere, and some of the continents, like "Lockyer Land," are sometimes seen white, as though covered with ice; while one island (marked on our map as Hall Island) has been seen so frequently thus, that it is very probable that here some mountain or table-land rises into the region of perpetual snow.

The cause of the red color of Mars has never been satisfactorily ascertained. Its atmosphere does not appear to be dark enough to produce such an effect, and perhaps as probable an explanation as any is one the suggestion of which is a little startling at first. It is that vegetation on Mars may be *red* instead of green! There is no intrinsic improbability in the idea, for we are even to-day unprepared to say with any certainty why vegetation is green here, and it is quite easy to conceive of atmospheric conditions which would make red the best absorber of the solar heat. Here, then, we find a planet on which we obtain many of the conditions of life which we know ourselves, and here, if anywhere in the system, we may allowably inquire for evidence of the presence of something like our own race: but though we may indulge in supposition, there is unfortunately no prospect that with any conceivable improvement in our telescopes we shall ever obtain anything like certainty. We cannot assert that there are any bounds to man's invention, or that science may not, by some means as unknown to us as the spectroscope was to our grandfathers, achieve what now seems impossible; but to our present knowledge no such means exist, though we are not forbidden to look at the ruddy planet with the feeling that it may hold possibilities more interesting to our humanity than all the wonders of the sun, and all the uninhabitable immensities of his other worlds.

Before we leave Mars, we may recall to the reader's memory the extraordinary verification of a statement made about it more than a hundred years ago. We shall have for a moment to leave the paths of science for those of pure fiction, for the words we are going to quote are those of no less a person than our old friend Captain Gulliver, who, after his adventures with the Lilliputians, went to a flying island inhabited largely by astronomers. If the reader will take down his copy of Swift, he will find in this voyage of Gulliver's to Laputa the following imaginary description of what its imaginary astronomers saw:—

"They have likewise discovered two lesser stars or satellites which revolve about Mars, whereof the innermost is distant from the centre of the primary planet exactly three of its diameters, and the outermost five; the former revolves in the space of ten hours, and the latter in twenty-one and a half."

Now, compare this passage, which was published in the year 1727, with the announcement in the scientific journals of August, 1877 (a hundred and fifty years after), that two moons did exist, and had just been discovered by Professor Hall, of

Washington, with the great telescope of which a drawing has been already given. The resemblance does not end even here, for Swift was right also in describing them as very near the planet and with very short periods, the actual distances being about one and a half and seven diameters, and the actual times about eight and thirty hours respectively, — distances and periods which, if not exactly those of Swift's description, agree with it in being less than any before known in the solar system. It is certain that there could not have been the smallest ground for a suspicion of their existence when "Gulliver's Travels" was written, and the coincidence — which is a pure coincidence — certainly approaches the miraculous. We can no longer, then, properly speak of "the snowy poles of moonless Mars," though it does still remain moonless to all but the most powerful telescopes in the world, for these bodies are the very smallest known in the system. They present no visible disks to measure, but look like the faintest of points of light, and their size is only to be guessed at from their brightness. Professor Pickering has carried on an interesting investigation of them. His method depended in part on getting holes of such smallness made in a plate of metal that the light coming through them would be comparable with that of the Martial moons in the telescope. It was found almost impossible to command the skill to make these holes small enough, though one of the artists employed had already distinguished himself by drilling a hole through a fine cambric needle *lengthwise,* so as to make a tiny steel tube of it. When the difficulty was at last overcome, the satellites were found to be less than ten miles in diameter, and a just impression both of their apparent size and light may be gathered from the statement that either roughly corresponds to that which would be given by a human hand held up at Washington, and viewed from Boston, Massachusetts, a distance of four hundred miles.

We approach now the only planet in which man is certainly known to exist, and which ought to have an interest for us superior to any which we have yet seen, for it is our own. We are voyagers on it through space, it has been said, as passengers on a ship, and many of us have never thought of any part of the vessel but the cabin where we are quartered. Some curious passengers (these are the geographers) have visited the steerage, and some (the geologists) have looked under the hatches, and yet it remains true that those in one part of our vessel know little, even now, of their fellow-voyagers in another. How much less, then, do most of us know of the ship itself, for we were all born on it, and have never once been off it to view it from the outside!

No world comes so near us in the aerial ocean as the moon; and if we desire to view our own earth as a planet, we may put ourselves in fancy in the place of a lunar observer. "Is it inhabited?" would probably be one of the first questions which he would ask, if he had the same interest in us that we have in him; and the answer to this would call out all the powers of the best telescopes such as we possess.

An old author, Fontenelle, has put in the mouth of an imaginary spectator a lively description of what would be visible in twenty-four hours to one looking down on the earth as it turned round beneath him. "I see passing under my eyes," he says, "all sorts of faces, — white and black and olive and brown. Now it's hats, and now turbans, now long locks and then shaven crowns; now come cities with steeples, next more with tall, crescent-capped minarets, then others with porcelain towers; now great desolate lands, now great oceans, then dreadful deserts, — in short, all the infinite variety the earth's surface bears." The truth is, however, that, looking at the earth from the moon, the largest moving animal, the whale or the elephant, would be utterly beyond our ken; and it is questionable whether the largest ship on the ocean would be visible, for the popular idea as to the magnifying power of great telescopes is exaggerated. It is probable that under any but extraordinary circumstances our lunar observer, with our best telescopes, could not bring the earth within less than an apparent distance of five hundred miles; and the reader may judge how large a moving object must be to be seen, much less recognized, by the naked eye at such a distance.

Of course, a chief interest of the supposition we are making lies in the fact that it will give us a measure of our own ability to discover evidences of life in the moon, if there are any such as exist here; and in this point of view it is worth while to repeat, that scarcely any temporary phenomenon due to human action could be even telescopically visible from the moon under the most favoring circumstances. An army such as Napoleon led to Russia might conceivably be visible if it moved in a dark solid column across the snow. It is barely possible that such a vessel as one of the largest ocean steamships might be seen, under very favorable circumstances, as a moving dot; and it is even quite probable that such a conflagration as the great fire of Chicago would be visible in the lunar telescope, as something like a reddish star on the night side of our planet; but this is all in this sort that could be discerned.

By making minute maps, or, still better, photographs, and comparing one year with another, much however might have been done by our lunar observer during this century. In its beginning, in comparison to the vast forests which then covered the North American continent, the cultivated fields along its eastern seaboard would have looked to him like a golden fringe bordering a broad mantle of green; but now he would see that the golden fringe has encroached upon the green farther back than the Mississippi, and he would gather his best evidence of change from the fact (surely a noteworthy one) that the people of the United States have altered the features of the world during the present century to a degree visible in another planet!

Our observer would probably be struck by the moving panorama of forests, lakes, continents, islands, and oceans, successively gliding through the field of view of his telescope as the earth revolved; but, travelling along beside it on his lunar station, he would hardly appreciate its actual flight through space, which is an easy thing to describe in figures, and a hard one to conceive. If we look up at the clock, and as we watch the pendulum recall that we have moved about nineteen miles at every beat, or in less than three minutes, over a distance greater than that which divides New York from Liverpool, we still probably but very imperfectly realize the fact that (dropping all metaphor) the earth is really a great projectile, heavier than the heaviest of her surface rocks, and traversing space with a velocity of over sixty times that of the cannon-ball. Even the firing of a great gun with a ball weighing one or two hundred pounds is, to the novice at least, a striking spectacle. The massive iron sphere is hoisted into the gun, the discharge comes, the ground trembles, and, as it seems, almost in the same instant, a jet rises where the ball has touched the water far away. The impression of immense velocity and of a resistless capacity of destruction in that flying mass is irresistible, and justifiable too: but what is this ball to that of the earth, which is a globe counting eight thousand miles in diameter, and weighing about six thousand millions of millions of millions of tons; which, if our cannon-ball were flying ahead a mile in advance of its track, would overtake it in less than the tenth part of a second; and which carries such a potency of latent destruction and death in this motion, that if it were possible instantly to arrest it, then, in that instant, "earth and all which it inherits would dissolve" and pass away in vapor?

Our turning sphere is moving through what seems to be all

but an infinite void, peopled only by wandering meteorites, and where warmth from any other source than the sun can scarcely be said to exist; for it is important to observe that whether the interior be molten or not, we get next to no heat from it. The cold of outer space can only be estimated in view of recent observations as at least four hundred degrees Fahrenheit below zero (mercury freezes at thirty-nine degrees below), and it is the sun which makes up the difference of all these lacking hundreds of degrees to us, but indirectly, and not in the way that we might naturally think, and have till very lately thought; for our atmosphere has a great deal to do with it beside the direct solar rays, allowing more to come in than to go out, until the temperature rises very much higher than it would were there no air here. Thus, since it is this power in the atmosphere of storing the heat which makes us live, no less than the sun's rays themselves, we see how the temperature of a planet may depend on considerations quite beside its distance from the sun; and when we discuss the possibility of life in other worlds, we shall do well to remember that Saturn may be possibly a warm world, and Mercury conceivably a cold one.

We used to be told that this atmosphere extended forty-five miles above us, but later observation proves its existence at a height of many times this; and a remarkable speculation, which Dr. Hunt strengthens with the great name of Newton, even contemplates it as extending in ever-increasing tenuity until it touches and merges in the atmosphere of other worlds.

But if we begin to talk of things new and old which interest us in our earth as a planet, it is hard to make an end. Still we may observe that it is the very familiarity of some of these which hinders us from seeing them as the wonders they really are. How has this familiarity, for instance, made commonplace to us not only the wonderful fact that the fields and forests, and the apparently endless plain of earth and ocean, are really parts

FIG. 65.—THE MOON.
(FROM A PHOTOGRAPH BY L. M. RUTHERFURD, 1873, PUBLISHED BY O. G. MASON.)

of a great globe which is turning round (for this rotation we all are familiar with), but the less appreciated miracle that we are all being hurled through space with an immensely greater speed than that of the rotation itself. It needs the vision of a poet to see this daily miracle with new eyes; and a great poet has described it for us, in words which may vivify our scientific conception. Let us recall the prologue to "Faust," where the archangels are praising the works of the Lord, and looking at the earth, not as we see it, but down on it, from heaven, as it passes by, and notice that it is precisely this miraculous swiftness, so insensible to us, which calls out an angel's wonder.

> "And swift and swift beyond conceiving
> The splendor of the world goes round,
> Day's Eden-brightness still relieving
> The awful Night's intense profound.
> The ocean tides in foam are breaking,
> Against the rocks' deep bases hurled,
> And both, the spheric race partaking,
> Eternal, swift, are onward whirled."[1]

So, indeed, might an angel see it and describe it!

We may have been already led to infer that there is a kind of evolution in the planets' life, which we may compare, by a not wholly fanciful analogy, to ours; for we have seen worlds growing into conditions which may fit them for habitability, and again other worlds where we may surmise, or may know, that life has come. To learn of at least one which has completed the analogy, by passing beyond this term to that where all life has ceased, we need only look on the moon.

The study of the moon's surface has been continued now from the time of Galileo, and of late years a whole class of competent observers has been devoted to it, so that astronomers engaged in other branches have oftener looked on this as a field

[1] Bayard Taylor's translation.

for occasional hours of recreation with the telescope than made it a constant study. I can recall one or two such hours in earlier observing days, when, seated alone under the overarching iron dome, the world below shut out, and the world above opened, the silence disturbed by no sound but the beating of the equatorial clock, and the great telescope itself directed to some hill or valley of the moon, I have been so lost in gazing that it seemed as though a look through this, the real magic tube, had indeed transported me to the surface of that strange alien world. Fortunately for us, the same spectacle has impressed others with more time to devote to it and more ability to render it, so that we not only have most elaborate maps of the moon for the professional astronomer, but abundance of paintings, drawings, and models, which reproduce the appearance of its surface as seen in powerful telescopes. None of the latter class deserves more attention than the beautiful studies of Messrs. Nasmyth and Carpenter, who prepared at great labor very elaborate and, in general, very faithful models of parts of its surface, and then had them photographed under the same illumination which fell on the original; and I wish to acknowledge here the special indebtedness of this part of what I have to lay before the reader to their work, from which the following illustrations are chiefly taken.

Let us remember that the moon is a little over twenty-one hundred miles in diameter; that it weighs, bulk for bulk, about two-thirds what the earth does, so that, in consequence of this and its smaller size, its total weight is only about one-eightieth of that of our globe; and that, the force of gravity at its surface being only one-sixth what it is here, eruptive explosions can send their products higher than in our volcanoes. Its area is between four and five times that of the United States, and its average distance is a little less than two hundred and forty thousand miles.

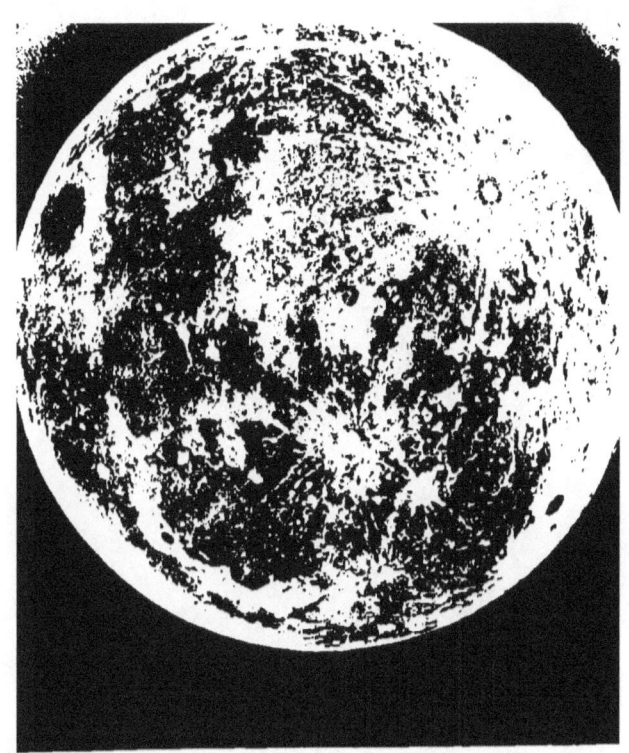

FIG. 66. — THE FULL MOON.

This is very little in comparison with the great spaces we have been traversing in imagination; but it is absolutely very large, and across it the valleys and mountains of this our nearest neighbor disappear, and present to the naked eye only the vague lights and shades known to us from childhood as "the man in the moon," and which were the puzzle of the ancient philosophers, who often explained them as reflections of the earth itself, sent back to us from the moon as from a mirror. It, at any rate, shows that the moon always turns the same face toward us, since we always see the same "man," and that there must be a back to the moon which we never behold at all; and, in fact, nearly half of this planet does remain forever hidden from human observation.

The "man in the moon" disappears when we are looking in a telescope, because we are then brought so near to details that the general features are lost; but he can be seen in any photograph of the full moon by viewing it at a sufficient distance, and making allowance for the fact that the contrasts of light and shade appear stronger in the photograph than they are in reality. If the small full moon given in Fig. 66, for instance, be looked at from across a room, the naked-eye view will be recovered, and its connection with the telescopic ones better made out. The best time for viewing the moon, however, is not at the full, but at the close of the first quarter; for then we see, as in this beautiful photograph (Fig. 65) by Mr. Rutherfurd, that the sunlight, falling slantingly on it, casts shadows which bring out all the details so that we can distinguish many of them even here, — this photograph, though much reduced, giving the reader a better view than Galileo obtained with his most powerful telescope. The large gray expanse in the lower part is the Mare Serenitatis, that on the left the Mare Crisium, and so on; these "seas," as they were called by the old observers, being no seas at all in reality, but extended plains which reflect less light than other portions, and which with higher powers show an irregular surface. Most of the names of the main features of the lunar surface were bestowed by the earlier observers in the infancy of the telescope, when her orb

> "Through optic glass the Tuscan artist 'viewed'
> At evening from the top of Fiesole
> Or in Valdarno, to descry new lands,
> Rivers, or mountains in her spotty globe."

Mountains there are, like the chain of the lunar Apennines, which the reader sees a little below the middle of the moon, and to the right of the Mare Serenitatis, and where a good telescope will show several thousand distinct summits. Apart from the mountain chains, however, the whole surface is visibly pitted with shallow, crater-like cavities, which vary from over a hundred miles in diameter to a few hundred yards or less, and which, we shall see later, are smaller sunken plains walled about with mountains or hills.

One of the most remarkable of these is Tycho, here seen on the photograph of the full moon (Fig. 66), from which radiating streaks go in all directions over the lunar surface. These streaks are a feature peculiar to the moon (at least we know of nothing to which they can be compared on the earth), for they run through mountain and valley for hundreds of miles without any apparent reference to the obstacles in their way, and it is clear that the cause is a deep-seated one. This cause is believed by our authors to be the fact that the moon was once a liquid sphere over which a hard crust formed, and that in subsequent time the expansion of the interior before solidification cracked the shell as we see. The annexed figure (Fig. 67) is furnished by them to illustrate their theory, and to show the effects of what they believe to be an analogous experiment, *in minimis*, to what Nature has performed on the grandest scale; for the photograph shows a glass globe actually cracked by the expansion of an

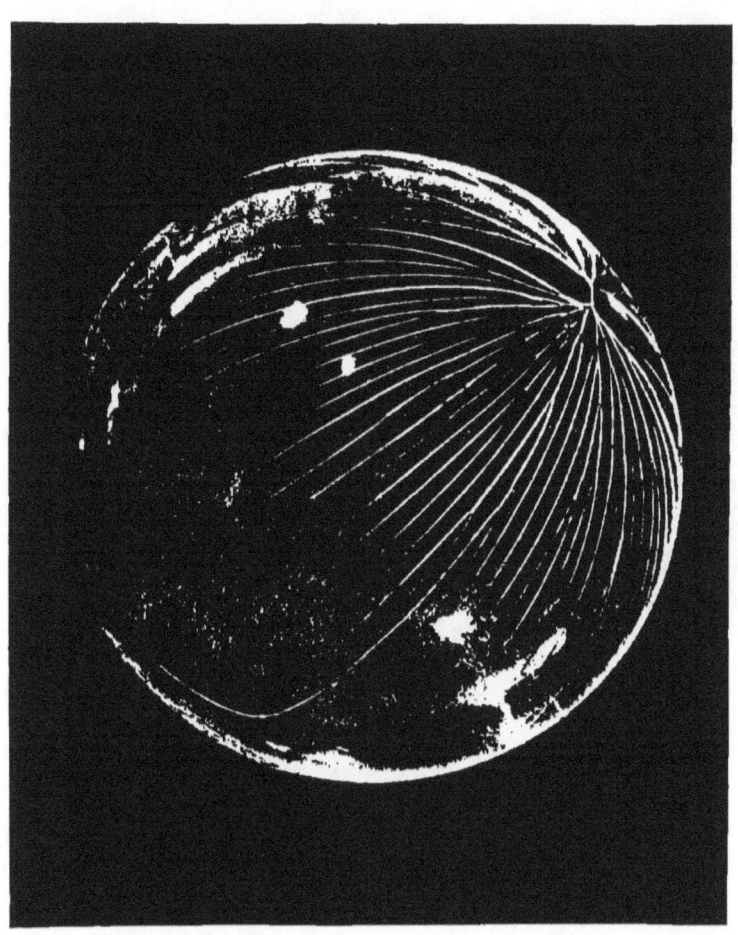

FIG. 67. — GLASS GLOBE, CRACKED.

enclosed fluid (in this case water), and the resemblance of the model to the photograph of the full moon on page 141 is certainly a very interesting one.

We are able to see from this, and from the multitude of craters shown even on the general view, where the whole face of our satellite is pit-marked, that eruptive action has been more prominent on the moon in ages past than on our own planet, and we are partly prepared for what we see when we begin to study it in detail.

We may select almost any part of the moon's surface for this nearer view, with the certainty of finding something interesting. Let us choose, for instance, on the photograph of the half-full moon (Fig. 65), the point near the lower part of the Terminator (as the line dividing light from darkness is called) where a minute sickle of light seems to invade the darkness, and let us apply in imagination the power of a large telescope to it. We are brought at once considerably within a thousand miles of the surface, over which we seem to be suspended, everything lying directly beneath us as in a bird's-eye view, and what we see is the remarkable scene shown in Fig. 68.

We have before us such a wealth of detail that the only trouble is to choose what to speak of where every point has something to demand attention, and we can only give here the briefest reference to the principal features. The most prominent of these is the great crater "Plato," which lies in the lower right-hand part of the cut. It will give the reader an idea of the scale of things to state that the diameter of its ring is about seventy miles; so that he will readily understand that the mountains surrounding it may average five to six thousand feet in height, as they do. The sun is shining from the left, and, being low, casts long shadows, so that the real forms of the mountains on one side are beautifully indicated by these shadows, where they fall on the floor of the crater. In the lower part of the mountain wall there has been a land-slide, as we see by the fragments that have rolled down into the plain, and of which a trace can be observed in our engraving. The whole is quite unlike most terrestrial craters, however, not only in its enormous size, but in its proportions; for the floor is not precipitous, but flat, or partaking of the general curvature of the lunar surface, which it sinks but little below. I have watched with interest in the telescope streaks and shades on the floor of Plato, not shown in our cut; for here some have suspected evidences of change, and fancied a faint greenish tint, as if due to vegetation, but it is probably fancy only. Notice the number of small craters around the big one, and everywhere on the plate, and then look at the amazingly rugged and tumbled mountain heaps on the left (the lunar Alps), cut directly through by a great valley (the valley of the Alps), which is at the bottom about six miles wide and extraordinarily flat, — flatter and smoother even than our engraving shows it, and looking as though a great engineering work, rather than an operation of Nature, were in question. Above this the mountain shadows are cast upon a wide plain, in which are both depressed pits with little mountain (or rather hill) rings about them, and extraordinary peaks, one of which, Pico (above the great crater), starts up abruptly to the height of eight thousand feet, a lunar Matterhorn.

If Mars were as near as the moon, we should see with the naked eye clouds passing over its face; and that we never do see these on the moon, even with the telescope, is itself a proof that none exist there. Now, this absence of clouds, or indeed of any evidence of moisture, is confirmed by every one of the nearer views like those we are here getting. We might return to this region with the telescope every month of our lives without finding one indication of vapor, of moisture, or even of air; and from a summit like Pico, could we ascend it, we should look out on a scene of such absolute desolation as probably no

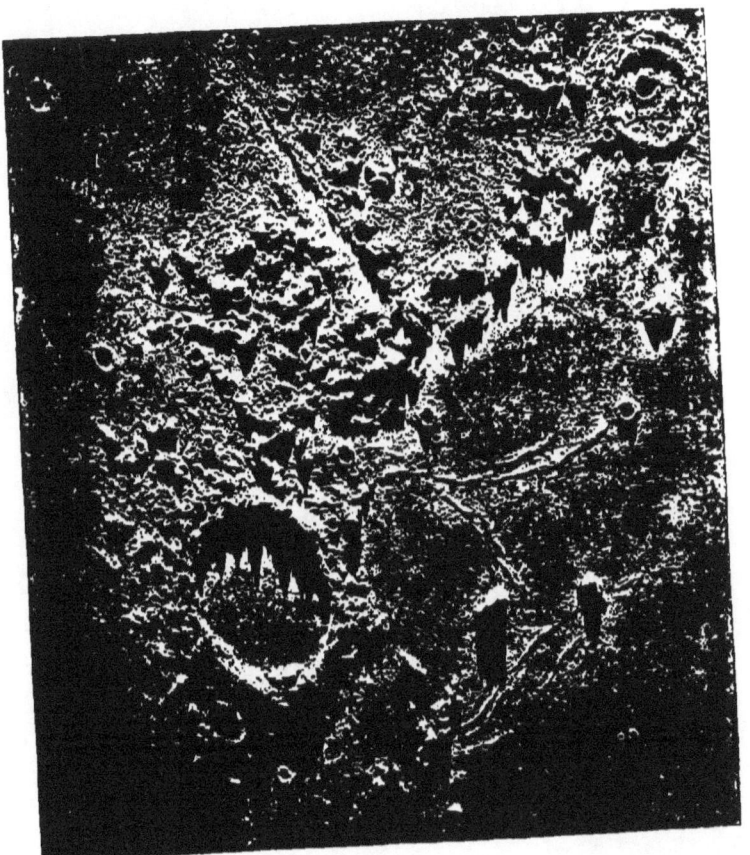

FIG. 58.—PLATO AND THE LUNAR ALPS.

earthly view could parallel. If, as is conceivable, these plains were once covered with verdure, and the abode of living creatures, verdure and life exist here no longer, and over all must be the silence of universal death. But we must leave it for another scene.

South of Plato extends for many hundred miles a great plain, which from its smoothness was thought by the ancient observers to be water, and was named by them the "Imbrian Sea," and this is bounded on the south and west by a range of mountains — the "lunar Apennines" (Fig. 69) — which are the most striking on our satellite. They are visible even with a spy-glass, looking then like bread-crumbs ranged upon a cloth, while with a greater power they grow larger and at the same time more chaotic. As we approach nearer, we see that they rise with a comparatively gradual slope, to fall abruptly, in a chain of precipices that may well be called tremendous, down to the plain below, across which their shadows are cast. Near their bases are some great craters of a somewhat different type from Plato, and our illustration represents an enlarged view of a part of this Apennine chain, of the great crater Archimedes, and of its companions Aristillus and Autolycus.

Our engraving will tell, more than any description, of the contrast of the tumbled mountain peaks with the level plain from which they spring, — a contrast for which we have scarcely a terrestrial parallel, though the rise of the Alps from the plains of Lombardy may suggest an inadequate one The Sierra Nevadas of California climb slowly up from the coast side, to descend in great precipices on the east, somewhat like this; but the country at their feet is irregular and broken, and their highest summits do not equal those before us, which rise to seventeen or eighteen thousand feet, and from one of which we should look out over such a scene of desolation as we can only imperfectly picture to ourselves from any experience of a terrestrial desert. The curvature of the moon's surface is so much greater than ours, that it would hide the spurs of hills which buttress the southern slopes of Archimedes, leaving only the walls of the great mountain ring visible in the extremest horizon, while between us and them would extend what some still maintain to have been the bed of an ancient lunar ocean, though assuredly no water exists there now.

Among the many fanciful theories to account for the forms of the ringed plains, one (and this is from a man of science whose ideas are always original) invokes the presence of water. According to it, these great plains were once ocean beds, and in them worked a coral insect, building up lunar "atolls" and ring-shaped submarine mountains, as the coral polyp does here. The highest summits of the great rings thus formed were then low islands, just "a-wash" with the waves of the ancient lunar sea, and, for aught we know, green with feathery palms. Then came (in the supposition in question) a time when the ocean dried up, and the mountains were left standing, as we see, in rings, after the cause of their formation was gone. If it be asked where the water went to, the answer is not very obvious on the old theories; but those who believe in them point to the extraordinary cracks in the soil, like those our engraving shows, as chasms and rents, by which the vanished seas, and perhaps also the vanished air, have been absorbed into the interior.

If there was indeed such an ancient ocean, it would have washed the very feet of the precipices on whose summits we are in imagination standing, and below us their recesses would have formed harbors which fancy might fill with commerce, and cities in which we might picture life and movement where all is now dead. It need hardly be said that no telescope has ever revealed their existence (if such ruins, indeed, there are), and it may be added that the opinion of geologists is, as a whole, unfavorable to the presence of water on the moon, even

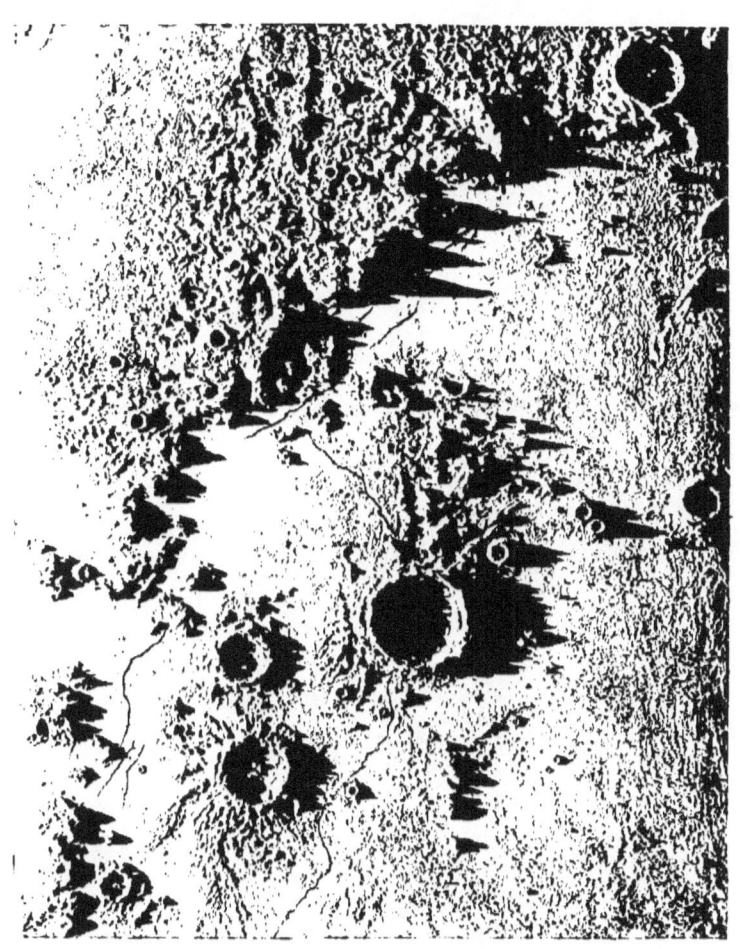

FIG. 69. — THE LUNAR APENNINES: ARCHIMEDES.

in the past, from the absence of any clear evidence of erosive action; but perhaps we are not yet entitled to speak on these points with certainty, and are not forbidden to believe that water may have existed here in the past by any absolute testimony to the contrary. The views of those who hold the larger portion of the lunar craters to have been volcanic in their formation are far more probable; and perhaps as simple an evidence of the presumption in their favor as we can give is directly to compare such a lunar region as this, the picture of which was made for us from a model, with a similar model made from some terrestrial volcanic region. Here (Fig. 70) is a photograph of such a modelled plan of the country round the Bay of Naples, showing the ancient crater of Vesuvius and its central cone, with other and smaller craters along the sea. Here, of course, we *know* that the forms originated in volcanic action, and a comparison of them with our moon-drawing is most interesting. To return to our Apennine region (Fig. 69), we must admit, however, when we consider the vast size of these things (Archimedes is fifty miles in diameter), that they are very different in proportion from our terrestrial craters, and that numbers of them present no central cone whatever; so that if some of them seem clearly eruptive, there are others to which we have great difficulties in making these volcanic theories apply. Let us look, for instance, at still another region (Fig. 71). It lies rather above the centre of the full moon, and may be recognized also on the Rutherfurd photograph; and it consists of the group of great ring-plains, three of which form prominent figures in our cut.

Ptolemy (the lower of these in the drawing) is an example of such a plain, whose diameter reaches to about one hundred and fifteen miles, so that it encloses an area of nearly eight thousand square miles (or about that of the State of Massachusetts), within which there is no central cone or point from which eruptive forces appear to have acted, except the smaller craters it encloses. On the south we see a pass in the mountain wall opening into the neighboring ring-plain of Alphonsus, which is only less in size; and south of this again is Arzachel, sixty-six miles in diameter, surrounded with terraced walls, rising in one place to a height greater than that of Mont Blanc, while the central cone is far lower. The whole of the region round about, though not the roughest on the moon, is rough and broken in a way beyond any parallel here, and which may speak for itself; but perhaps the most striking of the many curious features — at least the only one we can pause to examine — is what is called "The Railway," an almost perfectly straight line, on one side of which the ground has abruptly sunk, leaving the undisturbed part standing like a wall, and forming a "fault," as geologists call it. This is the most conspicuous example of its kind in the moon, but it is only one of many evidences that we are looking at a world whose geological history has been not wholly unlike our own. But the moon contains, as has been said, but the one-eightieth part of the mass of our globe, and has therefore cooled with much greater rapidity, so that it has not only gone through the epochs of our own past time, but has in all probability already undergone experiences which for us lie far in the future; and it is hardly less than justifiable language to say that we are beholding here in some respects what the face of our world may be when ages have passed away.

To see this more clearly, we may consider that in general we find that the early stages of cosmical life are characterized by great heat; a remark of the truth of which the sun itself furnishes the first and most obvious illustration. Then come periods which we appear to have seen exemplified in Jupiter, where the planet is surrounded by volumes of steam-like vapor, through which we may almost believe we recognize the dull glow of not yet extinguished fires; then times like those which

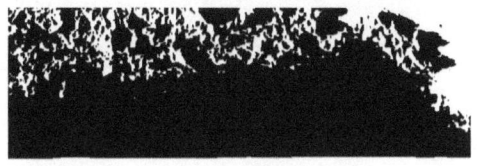

FIG. 70. — VESUVIUS AND NEIGHBORHOOD OF NAPLES.

our earth passed through before it became the abode of man; and then the times in which human history begins. But if this process of the gradual loss of heat go on indefinitely, we must yet come to still another era, when the planet has grown too cold to support life, as it was before too hot; and this condition, in the light of some very recent investigations, it seems probable we have now before us on the moon.

We have, it is true, been taught until very lately that the side of the moon turned sunward would grow hotter and hotter in the long lunar day, till it reached a temperature of two hundred to three hundred degrees Fahrenheit, and that in the equally long lunar night it would fall as much as this below zero. But the evidence which was supposed to support this conclusion as to the heat of the lunar day is not supported by recent experiments of the writer; and if these be trustworthy, certain facts appear to him to show that the temperature of the moon's surface, even under full perpetual sunshine, must be low, — and this because of the absence of air there to keep the stored sun-heat from being radiated away again into space.

As we ascend the highest terrestrial mountains, and get partly above our own protecting blanket of air, things do not grow hotter and hotter, but colder and colder; and it seems contrary to the teachings of common sense to believe that if we could ascend higher yet, where the air ceases altogether, we should not find that it grew colder still. But this last condition (of airlessness) is the one which does prevail beyond a doubt in the moon, on whose whole surface, then, there must be (unless there are sources of internal heat of which we know nothing) conditions of temperature which are an exaggeration of those we experience on the summit of a very lofty mountain, where we have the curious result that the skin may be burned under the solar rays, while we are shivering at the same time in what the thermometer shows is an arctic cold.

We have heard of this often; but a personal experience so impressed the fact on me that I will relate it for the benefit of the reader, who may wish to realize to himself the actual conditions which probably exist in the airless lunar mountains and plains we are looking at. He cannot go there; but he may go if he pleases, as I have done, to the waterless, shadeless waste which stretches at the eastern slope of the Sierra Nevadas (a chain almost as high and steep as the lunar Apennines), and live some part of July and August in this desert, where the thermometer rises occasionally to one hundred and ten degrees in the shade, and his face is tanned till it can tan no more, and he appears to himself to have experienced the utmost in this way that the sun can do.

The sky is cloudless, and the air so clear that all idea of the real distance and size of things is lost. The mountains, which rise in tremendous precipices above him, seem like moss-covered rocks close at hand, on the tops of which, here and there, a white cloth has been dropped; but the "moss" is great primeval forests, and the white cloths large isolated snow-fields, tantalizing the dweller in the burning desert with their delusive nearness. When I climbed the mountains, at an altitude of ten thousand feet I already found the coolness delicious, but at the same time (by the strange effect I have been speaking of) the skin began to burn, as though the seasoning in the desert counted for nothing at all; and as the air grew thinner and thinner while I mounted still higher and higher, though the thermometer fell, every part of the person exposed to the solar rays presented the appearance of a recent severe burn from an actual fire, — and a really severe burn it was, as I can testify, — and yet all the while around us, under this burning sun and cloudless sky, reigned a perpetual winter which made it hard to believe that torrid summer still lay below. The thinner the air, then, the colder it grows, even where we are exposed to the

FIG. 71. — PTOLEMY AND ARZACHEL.

sun, and the lower becomes the reading of the thermometer. Now, by means of suitable apparatus, it was sought by the writer to determine, while at this elevation of fifteen thousand feet, *how* great the fall of temperature would be if the thin air there could be removed altogether; and the result was that the thermometer would under such circumstances fall, at any rate, below zero in the full sunshine.

Of course, all this applies indirectly to the moon, above whose surface (if these inferences be correct) the mercury in the bulb of a thermometer would possibly freeze and never melt again during the lunar day (and still less during the lunar night), — a conclusion which has been reached through other means by Mr. Ericsson, — and whose surface itself can hardly be very greatly warmer. Other and direct measures of the lunar heat are still in progress while this is being written, but their probable result seems to be already indicated: it is that the moon's surface, even in perpetual sunshine, must be forever cold. Just how cold, is still doubtful; and it is not yet certain whether ice, if once formed there, could ever melt.

Here (Fig. 72) is one more scene from the almost unlimited field the lunar surface affords.

The most prominent things in the landscape before us are two fine craters (Mercator and Campanus), each over thirty miles in diameter; but we have chosen this scene for remark rather on account of the great crack or rift which is seen in the upper part, and which cuts through plain and mountain for a length of sixty miles. Such cracks are counted by hundreds on the moon, where they are to be seen almost everywhere; and other varieties, in fact, are visible on this same plate, but we will not stop to describe them. This one varies in width from an eighth of a mile to a mile; and though we cannot see to the bottom of it, others are known to be at least eight miles deep, and may be indefinitely deeper.

The edge of a cliff on the earth commonly gets weather-worn and rounded; but here the edge is sharp, so that a traveller along the lunar plains would come to the very brink of this tremendous chasm before he had any warning of its existence. It is usually thus with all such rifts; and the straightness and sharpness of the edge in these cases suggest the appearance of an ice-crack to the observer. I do not mean to assert that there is more than a superficial resemblance. I do not write as a geologist; but in view of what we have just been reading of the lunar cold, we may ask ourselves whether, if water ever did exist here, we should not expect to find perpetual ice, not necessarily glittering, but covered, perhaps, with the deposits of an air laden with the dust-products of later volcanic eruptions, or even covered in after ages, when the air has ceased from the moon, with the slow deposit of meteoric dust during millions of years of windless calm. What else can we think will become of the water on our own earth if it be destined to pass through such an experience as we seem to see prophesied in the condition of our dead satellite?

The reader must not understand me as saying that there is ice on the moon, — only that there is not improbably perpetual ice there now *if* there ever was water in past time; and he is not to suppose that to say this is in any way to deny what seems the strong evidence of the existence of volcanic action everywhere, for the two things may well have existed in successive ages of our satellite's past, or even have both existed together, like Hecla, within our own arctic snows; and if no sign of any still active lunar volcano has been discovered, we appear to read the traces of their presence in the past none the less clearly.

I remember that at one time, when living on the lonely upper lava-wastes of Mount Etna, which are pitted with little craters, I grew acquainted with so many a chasm and rent filled with these, that the dreary landscape appeared from above as if a bit

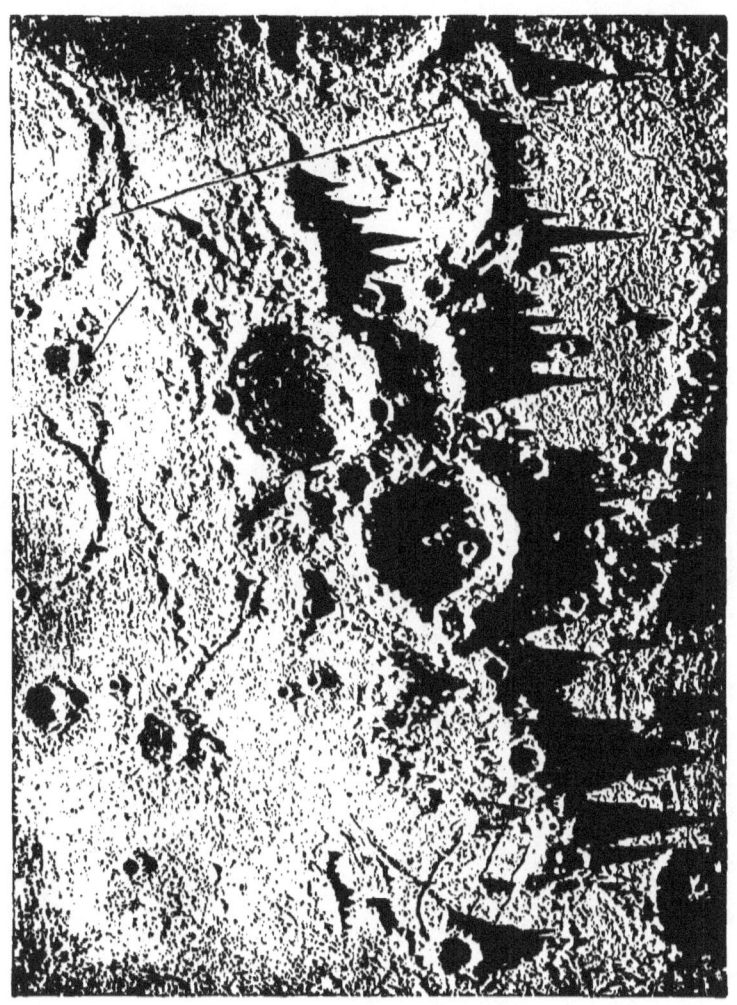

FIG. 72. — MERCATOR AND CAMPANUS.

of the surface of the moon I looked up at through the telescope had been brought down beside me.

I remember, too, that as I studied the sun there, and watched the volcanic outbursts on its surface, I felt that I possibly embraced in a threefold picture as many stages in the history of planetary existence, through all of which this eruptive action was an agent, — above in the primal energies of the sun; all around me in the great volcano, black and torn with the fires that still burn below, and whose smoke rose over me in the plume that floated high up from the central cone; and finally in this last stage in the moon, which hung there pale in the daylight sky, and across whose face the vapors of the great terrestrial volcano drifted, but on whose own surface the last fire was extinct.

We shall not get an adequate idea of it all, unless we add to our bird's-eye views one showing a chain of lunar mountains as they would appear to us if we saw them, as we do our own Alps or Apennines, from about their feet; and such a view Fig. 74 affords us. In the barren plain on the foreground are great rifts such as we have been looking at from above, and smaller craters, with their extinct cones; while beyond rise the mountains, ghastly white in the cold sunshine, their precipices crowned by no mountain fir or cedar, and softened by no intervening air to veil their nakedness.

If the reader has ever climbed one of the highest Alpine peaks, like those about Monte Rosa or the Matterhorn, and there waited for the dawn, he cannot but remember the sense of desolation and strangeness due to the utter absence of everything belonging to man or his works or his customary abode, above all which he is lifted into an upper world, so novel and, as it were, so unhuman in its features, that he is not likely to have forgotten his first impression of it; and this impression gives the nearest but still a feeble idea of what we see with the telescope in looking down on such a colorless scene, where too

168 THE NEW ASTRONOMY.

no water bubbles, no tree can sigh in the breeze, no bird can sing, — the home of silence.

But here, above it, hangs a world in the sky, which we should need to call in color to depict, for it is green and yellow

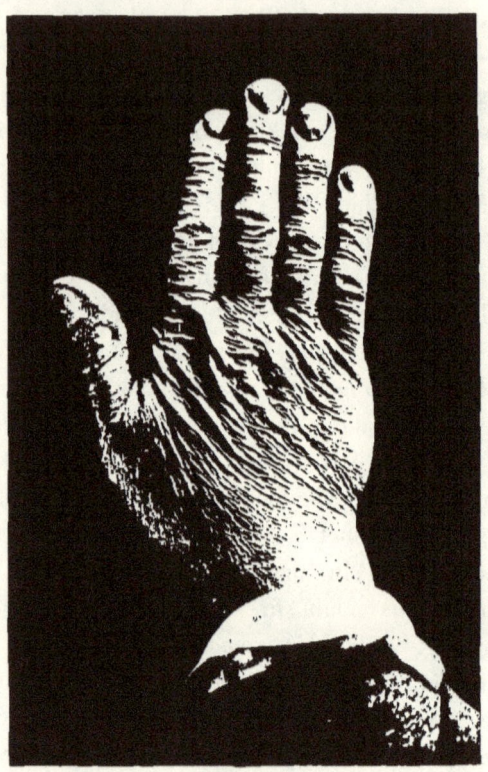

FIG. 73. — WITHERED HAND.

with the forests and the harvest-fields that overspread its continents, with emerald islands studding its gray oceans, over all of which sweep the clouds that bring the life-giving rain. It is our own world, which lights up the dreary lunar night, as the moon does ours.

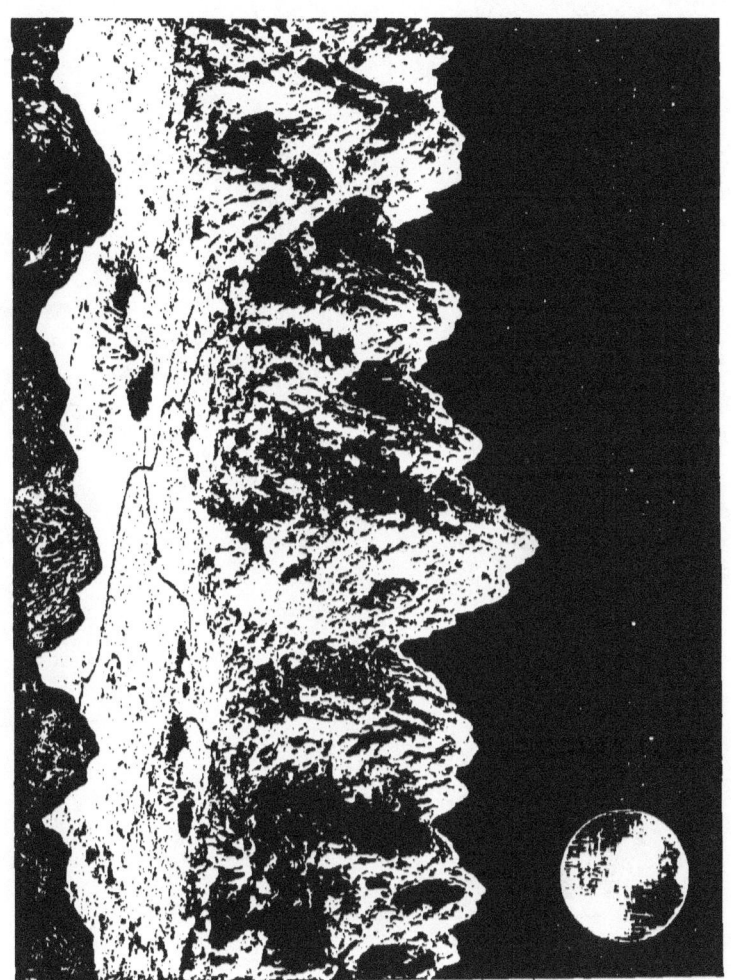

FIG. 74 — IDEAL LUNAR LANDSCAPE AND EARTH-SHINE.

The signs of age are on the moon. It seems pitted, torn, and rent by the past action of long-dead fires, till its surface is like a piece of porous cinder under the magnifying-glass, — a burnt-out cinder of a planet, which rolls through the void like a ruin of what has been; and, more significant still, this surface is wrinkled everywhere, till the analogy with an old and shrivelled face

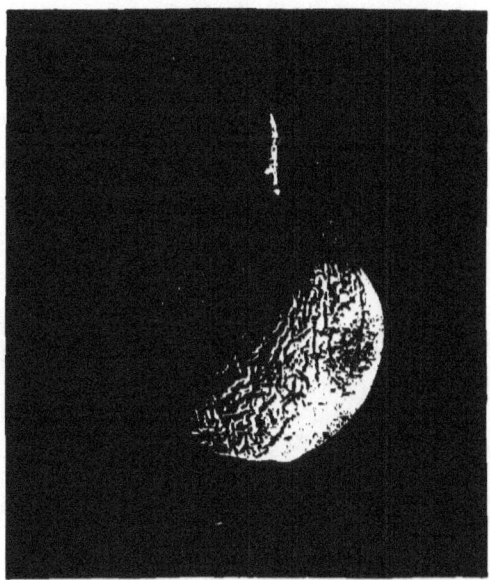

FIG. 75. — WITHERED APPLE.

or hand or fruit (Figs. 73 and 75), where the puckered skin is folded about a shrunken centre, forces itself on our attention, and suggests a common cause, — a something underlying the analogy, and making it more than a mere resemblance.

The moon, then, is dead; and if it ever was the home of a race like ours, that race is dead too. I have said that our New Astronomy modifies our view of the moral universe as well as of the physical one; nor do we need a more pregnant instance than in this before us. In these days of decay of old creeds of the eternal, it has been sought to satisfy man's yearning toward it by founding a new religion whose god is Humanity, and whose hope lies in the future existence of our own race, in whose collective being the individual who must die may fancy his aims and purpose perpetuated in an endless progress. But, alas for hopes looking to this alone! we are here brought to face the solemn thought that, like the individual, though at a little further date, Humanity itself may die!

Before we leave this dead world, let us take a last glance at one of its fairest scenes, — that which we obtain when looking at a portion on which the sun is rising, as in this view of Gassendi (Fig. 76), in which the dark part on our right is still the body of the moon, on which the sun has not yet risen. Its nearly level rays stretch elsewhere over a surface that is, in places, of a strangely smooth texture, contrasting with the ruggedness of the ordinary soil, which is here gathered into low plaits, that, with the texture we have spoken of, look

> "Like marrowy crapes of China silk,
> Or wrinkled skin on scalded milk,"

as they lie, soft and almost beautiful, in the growing light.

Where its first beams are kindling, the summits cast their shadows illimitedly over the darkening plains away on the right, until they melt away into the night, — a night which is not utterly black, for even here a subdued radiance comes from the earth-shine of our own world in the sky.

Let us leave here the desolation about us, happy that we can come back at will to that world, our own familiar dwelling, where the meadows are still green and the birds still sing, and where, better yet, still dwells our own kind, — surely the world, of all we have found in our wanderings, which we should ourselves have chosen to be our home.

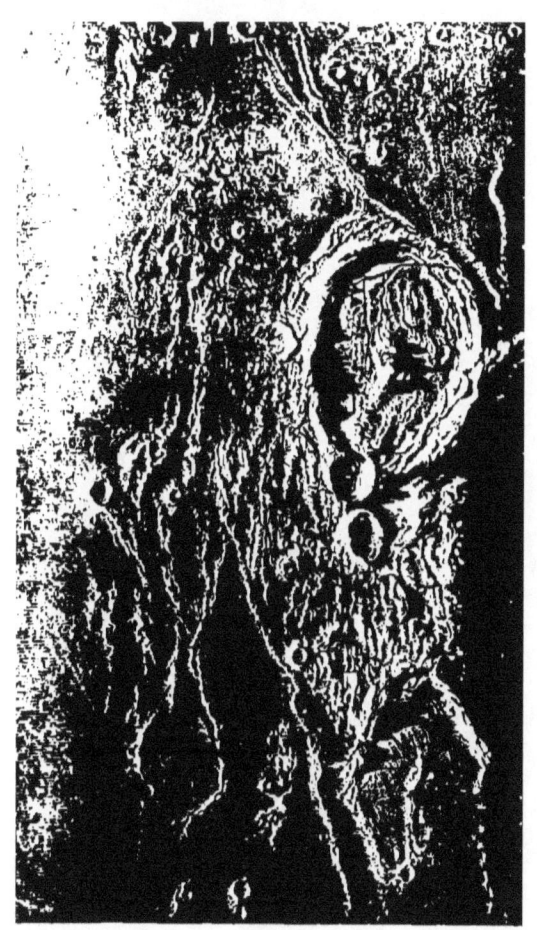

FIG. 76.— GASSENDI. NOV. 7, 1867.

VI.

METEORS.

WHAT is truth? What is fact, and what is fancy, even with regard to solid visible things that we may see and handle?

Among the many superstitions of the early world and credulous fancies of the Middle Ages, was the belief that great stones sometimes fell down out of heaven onto the earth.

Pliny has a story of such a black stone, big enough to load a chariot; the Mussulman still adores one at Mecca; and a mediæval emperor of Germany had a sword which was said to have been forged from one of these bolts shot out of the blue. But with the revival of learning, people came to know better! That stones should fall down from the sky was clearly, they thought, an absurdity; indeed, according to the learned opinion of that time, one would hardly ask a better instance of the difference between the realities which science recognized and the absurdities which it condemned than the fancy that such a thing could be. So at least the matter looked to the philosophers of the last century, who treated it much as they might treat certain alleged mental phenomena, for instance, if they were alive today, and at first refused to take any notice of these stories, when from time to time they still came to hand. When induced to give the matter consideration, they observed that all the conditions for scientific observation were violated by these bodies, since the wonder always happened at some far-off place or at some past time, and (suspicious circumstance!) the stones only fell in the presence of ignorant and unscientific witnesses, and never when scientific men were at hand to examine the facts. That there were many worthy, if ignorant, men who asserted that they had seen such stones fall, seen them with their very eyes, and held them in their own hands, was accounted for by the general love of the marvellous and by the ignorance of the common mind, unlearned in the conditions of scientific observation, and unguided by the great principle of the uniformity of the Laws of Nature.

Such a tone, of course, cannot be heard among us, who never hastily pronounce anything a departure from the "Laws of Nature," while uncertain that these can be separated from the laws of the fallible human mind, in which alone Nature is seen. But in the last century philosophers had not yet become humble, or scientific men diffident of the absoluteness of their own knowledge, and so it seemed that no amount of evidence was enough to gain an impartial hearing in the face of the settled belief that the atmosphere extended only a few miles above the earth's surface, and that the region beyond, whence alone such things could come, was an absolute void extending to the nearest planet.

It used to be supposed that we were absolutely isolated, not only from the stars but from other planets, by vast empty spaces extending from world to world, — regions altogether vacant except for some vagrant comet; but of late years we are growing to have new ideas on this subject, and not only to consider space as far from void or tenantless, but to admit, as a possibility at least, that there is a sort of continuity between our very earth's surface, the air above it, and all which lies beyond the blue overarching dome of our own sky. Our knowledge of the physical nature of the universe without has chiefly come from what the spectroscope, overleaping the space between us and

FIG. 77. — THE CAMP AT MOUNT WHITNEY.

(FROM "PROFESSIONAL PAPERS OF THE SIGNAL SERVICE," VOL. XV.)

the stars, has taught us of them; as a telegram might report to us the existence of a race across the ocean, without telling anything of what lay between. It would be a novel path to the stars, and to the intermediate regions whence these once mythical stones are now actually believed to come, if we could take the reader to them by a route which enabled us to note each step of a continuous journey from the earth's surface out into the unknown; but if we undertake to start upon it, he will understand that we must almost at the outset leave the ground of comparative certainty on which we have hitherto rested, and need to speak of things on this road which are still but probabilities, and even some which are little more than conjectures, before we get to the region of comparative certainty again, — a region which, strange to say, exists far away from us, while that of doubt lies close at hand, for we may be said without exaggeration to know more about Sirius than about the atmosphere a thousand miles above the earth's surface; indeed, it would be more just to say that we are sure not only of the existence but of the elements that compose a star, though a million of times as far off as the sun, while at the near point named we are not sure of so much as that the atmosphere exists at all.

To begin our outward journey in a literal sense, we might rise from the earth's surface some miles in a balloon, when we should find our progress stayed by the rarity of the air. Below us would be a gray cloud-ocean, through which we could see here and there the green earth beneath, while above us there would still be something in the apparently empty air, for if the sun has just set it will still be *light* all round us. Something then, in a cloudless sky, still exists to reflect the rays towards us, and this something is made up of separately invisible specks of dust and vapor, but very largely of actual dust, which probably forms the nucleus of each mist-particle. That discrete matter of some kind exists here has long been recognized from the phenomena of twilight; but it is, I think, only recently that we are coming to admit that a shell of actual solid particles in the form of dust probably encloses the whole globe, up to far above the highest clouds.

In 1881 the writer had occasion to conduct a scientific expedition to the highest point in the territories of the United States, on one of the summits of the Sierra Nevadas of Southern California, which rise even above the Rocky Mountains.

The illustration on page 177 represents the camp occupied by this party below the summit, where the tents, which look as if in the bottom of a valley, are yet really above the highest zone of vegetation, and at an altitude of nearly twelve thousand feet.

Still above these rise the precipices of barren rock seen in the background, their very bases far above the highest visible dust-clouds, which overspread like a sea the deserts at the mountain's foot, — precipices which when scaled lift the observer into what is, perhaps, the clearest and purest air to be found in the world. It will be seen from the mere looks of the landscape that we are far away here from ordinary sources of contamination in the atmosphere. Yet even above here on the highest peak, where we felt as if standing on the roof of the continent and elevated into the great aerial currents of the globe, the telescope showed particles of dust in the air, which the geologists deemed to have probably formed part of the soil of China and to have been borne across the Pacific, but which also, as we shall see later, may owe something to the mysterious source of the phenomena already alluded to.

It is far from being indifferent to us that the dust is there; for, to mention nothing else, without it, it would be night till the sunrise, and black night again as soon as the sun's edge disappeared below the horizon. The morning and the evening twilight, which in northern latitudes increase our average time of light by some hours, and add very materially to the actual days of man's life, are probably due almost wholly to particles scarcely visible in the microscope, and to the presence of such atoms, smaller than the very motes ordinarily seen in the sunbeam, which, as Mr. Aitken has shown, fill the air we breathe, — so minute and remote are the causes on which the habits of life depend.

Before we can see that a part of this impalpable, invisible dust is also perhaps a link between our world and other members of the solar system, we must ask how it gets into the atmosphere. Is it blown up from the earth, or does it fall down out of the miscalled "void" of space?

If we cast a handful of dust into the air, it will not mount far above the hand unless we set the air in motion with it, as in ascending smoke-currents; and the greatest explosions we can artificially produce, hurl their finer products but a few hundred feet at most from the soil. Utterly different are the forces of Nature. We have on page 183 a reproduction from a photograph of an eruption of Vesuvius, — a mere toy-volcano compared to Etna or Hecla. But observe the smoke-cloud which rises high in the sunshine, looking solid as the rounded snows of an Alp, while the cities and the sea below are in the shadow. The smoke that mounts from the foreground, where the burning lava-streams are pouring over the surface and firing the woods, is of another kind from that rolling high above. *This* comes from within the mountain, and is composed of clouds of steam mingled with myriads of dust-particles from the comminuted products of the earth's interior; and we can see ourselves that it is borne away on a level, miles high in the upper air.

But what is this to the eruption of Sumbawa or Krakatao? The latter occurred in 1883, and it will be remembered that the air-wave started by the explosion was felt around the globe, and that, probably owing to the dust and water-vapor blown into the atmosphere, the sunsets even in America became of that extraordinary crimson we all remember three years ago; and coincidently, that dim reddish halo made its appearance about the sun, the world over, which is hardly yet gone.[1] Very careful estimates of the amount of ashes ejected have been made; and though most of the heavier particles are known to have fallen into the sea within a few miles, a certain portion — the lightest — was probably carried by the explosion far above the lower strata of the atmosphere, to descend so slowly that some of it may still be there. Of this lighter class the most careful estimates must be vague; but according to the report of the official investigation by the Dutch Government, that which remained floating is something enormous. An idea of its amount may be gained by supposing these impalpable and invisible particles to condense again from the upper sky, and to pour down on the highest edifice in the world, the Washington Monument. If the dust were allowed to spread out on all sides, till the pyramidal slope was so flat as to be permanent, the capstone of the monument would not only be buried before the supply was exhausted, but buried as far below the surface as that pinnacle is now above it.

Of the explosive suddenness with which the mass was hurled, we can judge something (comparing small things with great) by the explosion of dynamite.

It happened once that the writer was standing by a car in which some railway porters were lifting boxes. At that moment came an almost indescribable sound, for it was literally stunning, though close and sharp as the crack of a whip in one's hand, and yet louder than the nearest thunder-clap. The men leaped from the car, thinking that one of the boxes had exploded between them; but the boxes were intact, and we saw what seemed a pillar of dust rising above the roof of the station, hundreds of yards away. When we hurried through the building,

[1] In January, 1887.

FIG. 78. — VESUVIUS DURING AN ERUPTION.

we found nothing on the other side but a bare plain, extending over a mile, and beyond this the actual scene of the explosion that had seemed to be at our feet. There had been there, a few minutes before, extensive buildings and shops belonging to the railroad, and sidings on which cars were standing, two of which, loaded with dynamite, had exploded.

Where they *had* been was a crater-like depression in the earth, some rods in diameter; the nearest buildings, great solid structures of brick and stone, had vanished, and the more distant wooden ones and the remoter lines of freight-cars on the sidetracks presented a curious sight, for they were not shattered so much as bent and leaning every way, as though they had been built of pasteboard, like card-houses, and had half yielded to some gigantic puff of breath. All that the explosion had shot skyward had settled to earth or blown away before we got in sight of the scene, which was just as quiet as it had been a minute before. It was like one of the changes of a dream.

Now, it is of some concern to us to know that the earth holds within itself similar forces, on an incomparably greater scale. For instance, the explosion which occurred at Krakatao, at five minutes past ten, on the 27th of August, 1883, according to official evidence, was heard at a distance of eighteen hundred miles, and the puff of its air-wave injured dwellings two hundred miles distant, and, we repeat, carried into the highest regions of the atmosphere and around the world matter which it is at least possible still affects the aspect of the sun to-day from New York or Chicago.

Do not the great flames which we have seen shot out from the sun at the rate of hundreds of miles a second, the immense and sudden perturbations in the atmosphere of Jupiter, and the scarred surface of the moon, seem to be evidences of analogous phenomena, common to the whole solar system, not wholly unconnected with those of earthquakes, and which we can still study in the active volcanoes of the earth?

If the explosion of gunpowder can hurl a cannon-shot three or four miles into the air, how far might the explosion of Krakatao cast its fragments? At first we might think there must be some proportionality between the volume of the explosion and the distance, but this is not necessarily so. Apart from the resistance of the air, it is a question of the velocity with which the thing is shot upward, rather than the size of the gun, or the size of the thing itself, and with a sufficient velocity the projectile would never fall back again. "What goes up must come down," is, like most popular maxims, true only within the limits of ordinary experience; and even were there nothing else in the universe to attract it, and though the earth's attraction extend to infinity, so that the body would never escape from it, it is yet quite certain that it would, with a certain initial velocity (very moderate in comparison with that of the planet itself), go up and *never* come back; while under other and possible conditions it might voyage out into space on a comet-like orbit, and be brought back to the earth, perhaps in after ages, when the original explosion had passed out of memory or tradition. But because all this is possible, it does not follow that it is necessarily true; and if the reader ask why he should then be invited to consider such suppositions at all, we repeat that in our journey outward, before we come to the stars, of which we know something, we pass through a region of which we know almost nothing; and this region, which is peopled by the subjects of conjecture, is the scene, if not the source, of the marvel of the falling stones, concerning which the last century was so incredulous, but for which we can, aided by what has just been said, now see at least a possible cause, and to which we now return.

Stories of falling stones, then, kept arising from time to time during the last century as they had always done, and philosophers kept on disbelieving them as they had always done, till an event occurred which suddenly changed scientific opinion to compulsory belief.

On the 26th of April, 1803, there fell, not in some far-off part of the world, but in France, not one alone, but many thousand stones, over an area of some miles, accompanied with noises like the discharge of artillery. A committee of scientific men visited the spot on the part of the French Institute, and brought back not only the testimony of scores of witnesses or auditors, but the stones themselves. Soon after stones fell in Connecticut, and here and elsewhere, as soon as men were prepared to believe, they found evidence multiplied; and such falls, it is now admitted, though rare in any single district, are of what may be called frequent occurrence as regards the world at large, — for, taking land and sea together, the annual stone-falls are probably to be counted by hundreds.

It was early noticed that these stones consisted either of a peculiar alloy of iron, or of minerals of volcanic origin, or both; and the first hypothesis was that they had just been shot out from terrestrial volcanoes. As they were however found, as in the case of the Connecticut meteorite, thousands of miles from any active volcanoes, and were seen to fall, not vertically down, but as if shot horizontally overhead, this view was abandoned. Next the idea was suggested that they were coming from volcanoes in the moon; and though this had little to recommend it, it was adopted in default of a better, and entertained down to a comparatively very recent period. These stones are now collected in museums, where any one may see them, and are to be had of the dealers in such articles by any who wish to buy them. They are coming to have such a considerable money value that, in one case at least, a lawsuit has been instituted for their possession between the finder, who had picked the stones up on ground leased to him, and claimed them under the tenant's right to wild game, and his landlord, who thought they were his as part of the real estate.

Leaving the decision of this novel law-point to the lawyers, let us notice some facts now well established.

The fall is usually preceded by a thundering sound, sometimes followed or accompanied by a peculiar noise described as like that of a flock of ducks rising from the water. The principal sound is often, however, far louder than any thunder, and sometimes of stunning violence. At night this is accompanied by a blaze of lightning-like suddenness and whiteness, and the stones commonly do not fall vertically, but as if shot from a cannon at long range. They are usually burning hot, but in at least one authenticated instance one was so intensely cold that it could not be handled. They are of all sizes, from tons to ounces, comparatively few, however, exceeding a hundred-weight, and they are oftenest of a rounded form, or looking like pieces of what was originally round, and usually wholly or partly covered with a glaze formed of the fused substance itself. If we slowly heat a lump of loaf sugar all through, it will form a pasty mass, while we may also hold it without inconvenience in our fingers to the gas-flame a few seconds, when it will be melted only on the side next the sudden heat, and rounded by the melting. The sharp contrast of the melted and the rough side is something like that of the meteorites; and just as the sugar does not burn the hand, though close to where it is brought suddenly to a melting heat, a mass of ironstone may be suddenly heated on the surface, while it remains cold on the inside. But, however it got there, the stone undoubtedly comes from the intensely cold spaces above the upper air; and what is the source of such a heat that it is melted in the cold air, and in a few seconds?

Everybody has noticed that if we move a fan gently, the air parts before it with little effort, while, when we try to fan vio-

FIG. 79. — METEORS OBSERVED NOV. 13 AND 14, 1868, BETWEEN MIDNIGHT AND FIVE O'CLOCK, A. M.

lently, the same air is felt to react; yet if we go on to say that if the motion is still more violent the atmosphere will resist like a solid, against which the fan, if made of iron, would break in pieces, this may seem to some an unexpected property of the "nimble" air through which we move daily. Yet this is the case; and if the motion is only so quick that the air cannot get out of the way, a body hurled against it will rise in temperature like a shot striking an armor-plate. It is all a question of speed, and that of the meteorite is known to be immense. One has been seen to fly over this country from the Mississippi to the Atlantic in an inappreciably short time, probably in less than two minutes; and though at a presumable height of over fifty miles, the velocity with which it shot by gave every one the impression that it went just above his head, and some witnesses of the unexpected apparition looked the next day to see if it had struck their chimneys. The heat developed by arrested motion in the case of a mass of iron moving twenty miles a second can be calculated, and is found to be much more than enough, not only to melt it, but to turn it into vapor; though what probably does happen is, according to Professor Newton, that the melted surface-portions are wiped away by the pressure of the air and volatilized to form the luminous train, the interior remaining cold, until the difference of temperature causes a fracture, when the stone breaks and pieces fall, — some of them at red-hot heat, some of them possibly at the temperature of outer space, or far below that of freezing mercury.

Where do these stones come from? What made them? The answer is not yet complete; but if a part of the riddle is already yielding to patience, it is worthy of note, as an instance of the connection of the sciences, that the first help to the solution of this astronomical enigma came from the chemists and the geologists.

The earliest step in the study, which has now been going on

for many years, was to analyze the meteorite, and the first result was that it contained no elements not found on this planet. The next was that, though none of these elements were unknown, they were not combined as we see them in the minerals we dig from the earth. Next it was found that the combinations, if unfamiliar at the earth's surface and nowhere reproduced exactly, were at least very like such as existed down beneath it, in lower strata, as far as we can judge by specimens of the earth's interior cast up from volcanoes. Later, a resemblance was recognized in the elements of the meteorites to those found by the spectroscope in shooting stars, though the spectroscopic observation of the latter is too difficult to have even yet proceeded very far. And now, within the last few years, we seem to be coming near to a surprising solution.

It has now been shown that meteoric stones sometimes contain pieces of essentially different rocks fused together, and pieces of detritus, — the wearing down of older rocks. Thus, as we know that sandstone is made of compacted sand, and sand itself was in some still earlier time part of rocks worn down by friction, — when it is shown, as it has been by M. Meunier, that a sandstone penetrated by metallic threads (like some of our terrestrial formations) has come to us in a meteorite, the conclusion that these stones may be part of some old world is one that, however startling, we cannot refuse at least to consider. According to this view, there may have been a considerable planet near the earth, which, having reached the last stage of planetary existence shown in the case of our present moon, went one step further, — went, that is, out of existence altogether, by literal breaking up and final disappearance. We have seen the actual moon scarred and torn in every direction, and are asked to admit the possibility that a continuance of the process on a similar body has broken it up into the fragments that come to us. We do not say that this is the case, but that (as regards the origin of some of the meteorites at least) we cannot at present disprove it. We may, at any rate, present to the novelist seeking a new *motif* that of a meteorite bringing to us the story of a lost race, in some fragment of art or architecture of its lost world!

We are not driven to this world-shattering hypothesis by the absence of others, for we may admit these to be fragments of a larger body without necessarily concluding that it was a world like ours, or, even if it were, that the world which sent them to us is destroyed. In view of what we have been learning of the tremendous explosive forces we see in action on the sun and probably on other planets, and even in terrestrial volcanoes to-day, it is certainly conceivable that some of these stones may have been ejected by some such process from any sun, or star, or world we see. The reader is already prepared for the suggestion that part of them may be the product of terrestrial volcanoes in early epochs, when our planet was yet glowing sunlike with its proper heat, and the forces of Nature were more active; and that these errant children of mother earth's youth, after circulating in lengthened orbits, are coming back to her in her age.

Do not let us, however, forget that these are mostly speculations only, and perhaps the part of wisdom is not to speculate at all till we learn more facts; but are not the facts themselves as extraordinary as any invention of fancy?

Although it is true that the existence of the connection between shooting stars and meteorites lacks some links in the chain of proof, we may very safely consider them together; and if we wish to know what the New Astronomy has done for us in this field, we should take up some treatise on astronomy of the last century. We turn in one to the subject of falling stars, and find that "this species of Star is only a light Exhalation, almost wholly sulphurous, which is inflamed in the free Air much after the same manner as Thunder in a Cloud by the blowing of the Winds." That the present opinion is different, we shall shortly notice.

All of us have seen shooting stars, and they are indeed something probably as old as this world, and have left their record in mythology as well as in history. According to Moslem tradition, the evil genii are accustomed to fly at night up to the confines of heaven in order to overhear the conversation of the angels, and the shooting stars are the fiery arrows hurled by the latter at their lurking foes, with so good an aim that we are told that for every falling star we may be sure that there is one spirit of evil the less in the world. The scientific view of them, however, if not so consolatory, is perhaps more instructive, and we shall here give most attention to the latter.

To begin with, there have been observed in history certain times when shooting stars were unusually numerous. The night when King Ibrahim Ben Ahmed died, in October, 902, was noted by the Arabians as remarkable in this way; and it has frequently been observed since, that, though we can always see some of these meteors nightly, there are at intervals very special displays of them. The most notable modern one was on Nov. 13, 1833, and this was visible over much of the North American continent, forming a spectacle of terrifying grandeur. An eye-witness in South Carolina wrote: —

"I was suddenly awakened by the most distressing cries that ever fell on my ears. Shrieks of horror and cries for mercy I could hear from most of the negroes of the three plantations, amounting in all to about six hundred or eight hundred. While earnestly listening for the cause I heard a faint voice near the door, calling my name. I arose, and, taking my sword, stood at the door. At this moment I heard the same voice still beseeching me to rise, and saying, 'O my God, the world is on fire!' I then opened the door, and it is difficult to say which excited me the most — the awfulness of the scene, or the distressed cries of the negroes. Upwards of one hundred lay prostrate on the ground,— some speechless and some with the bitterest cries, but with their hands raised, imploring God to save the world and them. The scene was truly awful; for never did rain fall much thicker than the meteors fell toward the earth; east, west, north, and south, it was the same."

The illustration on page 189 does not exaggerate the number of the fiery flashes at such a time, though the zigzag course which is observed in some is hardly so common as it here appears.

When it was noted that the same date, November 13th, had been distinguished by star-showers in 1831 and 1832, and that the great shower observed by Humboldt in 1799 was on this day, the phenomenon was traced back and found to present itself about every thirty-three years, the tendency being to a little delay on each return; so that Professor Newton and others have found it possible with this clew to discover in early Arabic and other mediæval chronicles, and in later writers, descriptions which, fitted together, make a tolerably continuous record of this thirty-three-year shower, beginning with that of King Ibrahim already alluded to. The shower appeared again in November, 1867 and 1868, with less display, but with sufficient brilliance to make the writer well remember the watch through the night, and the count of the flying stars, his most lively recollection being of their occasional colors, which in exceptional cases ranged from full crimson to a vivid green. The count on this night was very great, but the number which enter the earth's atmosphere even ordinarily is most surprising; for, though any single observer may note only a few in his own horizon, yet, taking the world over, at least ten millions appear every night, and on these special occasions very many more. This November shower comes always from a particular quarter of the sky, that occupied by the constellation Leo, but there are others, such as that of August 10th (which is annual), in which the "stars" seem to be shot at us from the constellation Perseus; and each of the numerous groups of star-showers is now known by the name of the constellation whence it seems to come, so that we have *Perseids* on August 10th, *Geminids* on December 12th, *Lyrids*, April 20th, and so on.

The great November shower, which is coming once more in this century, and which every reader may hope to see toward 1899, is of particular interest to us as the first whose movements were subjected to analysis; for it has been shown by the labors of Professor Newton, of Yale, and Adams, of Cambridge, that these shooting stars are bodies moving around the sun in an orbit which is completed in about thirty-three years. It is quite certain, too, that they are not exhalations from the earth's atmosphere, but little solids, invisible till they shine out by the light produced by their own fusion. Each, then, moves on its own track, but the general direction of all the tracks is the same; and though some of them may conceivably be solidified gases, we should think of them not as gaseous in form, but as solid shot, of the average size of something like a cherry, or perhaps even of a cherry-stone, yet each an independent planetoid, flying with a hundred times the speed of a rifle-bullet on its separate way as far out as the orbit of Uranus; coming back three times in a century to about the earth's distance from the sun, and repeating this march forever, unless it happen to strike the atmosphere of the earth itself, when there comes a sudden flash of fire from the contact, and the distinct existence of the little body, which may have lasted for hundreds of thousands of years, is ended in a second.

If the reader will admit so rough a simile, we may compare such a flight of these bodies to a thin swarm of swift-flying birds — thin, but yet immensely long, so as to be, in spite of the rapid motion, several years in passing a given point, and whose line of flight is cut across by us on the 13th of November, when the earth passes through it. We are only there on that day, and can only see it then; but the swarm is years in all getting by, and so we may pass into successive portions of it on the anniversary of the same day for years to come. The stars appear to shoot from Leo, only because that constellation is in the line of their flight when we look up to it, just as an interminable train of parallel flying birds would appear to come from some definite point on the horizon.

We can often see the flashes of meteors at over a hundred miles, and though at times they may seem to come thick as flakes of falling snow, it is probable, according to Professor Newton, that even in a "shower" each tiny planetoid is more than ten miles from its nearest neighbor, while on the average it is reckoned that we may consider that each little body, though possibly no larger than a pea, is over two hundred miles from its neighbor, or that to each such grain there are nearly ten million cubic miles of void space. Their velocity as compounded with that of the earth is enormous, sometimes forty to fifty miles per second (according to a recent but unproved theory of Mr. Denning, it would be much greater), and it is this enormous rate of progress that affords the semblance of an abundant fall of rain, notwithstanding the distance at which one drop follows another. It is only from their light that we are able to form a rough estimate of their average size, which is, as we have seen, extremely small; but, from their great number, the total weight they add to the earth daily may possibly be a hundred tons, probably not very much more. As they are as a rule entirely dissipated in the upper air, often at a height of from fifty to seventy miles, it follows that many tons of the finest pulverized and gaseous matter are shot into the earth's atmosphere every twenty-four hours from outer space, so that here is an independent and constant supply of dust, which we may expect to find coming down from far above the highest clouds.

Now, when the reader sees the flash of a shooting star, he may, if he please, think of the way the imagination of the East accounts for it, or he may look at what science has given him instead. In the latter case he will know that a light which flashed and faded almost together came from some strange little entity which had been traversing cold and vacant space for untold years, to perish in a moment of more than fiery heat; an enigma whose whole secret is unknown, but of which, during that instant flash, the spectroscope caught a part, and found evidence of the identity of some of its constituents with those of the observer's own body.

VII.

COMETS.

OF comets, the Old Astronomy knew that they came to the sun from great distances in all directions, and in calculable orbits; but as to *what* they were, this, even in the childhood of those of us who are middle-aged, was as little known as to the centuries during which they still from their horrid heads shook pestilence and war. We do not know even now by any means exactly what they are, for enough yet remains to be learned about them still to give their whole study the attraction which belongs to the unknown; and yet we have learned so much, and in a way which to our grandfathers would have been so unexpected, connecting together the comet, the shooting star, and the meteorite, that the astronomer who perhaps speaks with most authority about these to-day was able, not long ago, in beginning a lecture, to state that he held in his hand what had been a part of a comet; and what he held was, not something half vaporous or gaseous, as we might suppose from our old associations, but a curious stone like this on page 203, which, with others, had fallen from the sky in Iowa, a flashing prodigy, to the terror of barking dogs, shying horses, and fearful men, followed by clouds of smoke and vapor, and explosions that shook the houses like an earthquake, and "hollow bellowings and rattling sounds, mingled with clang and clash and roar," as an auditor described it. It is only a fragment of a larger stone which may have weighed tons. It looks inoffensive enough now, and its appearance affords no hint of the commotion it caused in a peaceable neighborhood only ten years ago. But what, it may be asked, is the connection between such things and comets?

To answer this, let us recall the statement that the orbit of the November meteor swarm has been computed; which means that those flying bodies have been found to come only from one particular quarter out of all possible quarters, at one particular angle out of all possible angles, at one particular velocity out of all possible velocities, and so on; so that the chances are endless against mere accident having produced another body which agreed in all these particulars, and others besides. Now, in 1867 the remarkable fact was established that a comet seen in the previous year (Comet 1, 1866) had the same orbit as the meteoroids, which implies, as we have just seen, that the comet and the meteors were in some way closely related.

The paths of the August meteors and of the Lyrids also have both been found to agree closely with those of known comets, and there is other evidence which not only connects the comets and the shooting stars, and makes it probable that the latter are due to some disintegration of the former, but even looks as though the process were still going on. And now with this in mind we may, perhaps, look at these drawings with more interest.

We have all seen a comet, and we have all felt, perhaps, something of the awe which is called up by the thought of its immensity and its rush through space like a runaway star. Its head is commonly like a small luminous point, from which usually grows as it approaches the sun a relatively enormous brush or tail of pale light, which has sometimes been seen to stretch across the whole sky from zenith to horizon. It is useless to look only along the ecliptic road for a comet's coming; rather may we expect to see it rushing down from above, or up from below, sometimes with a speed which is possibly greater

FIG. 80. — COMET OF DONATI, SEPT. 16, 1858.[1]

[1] The five engravings of the Comet of Donati are from "Annals of the Astronomical Observatory of Harvard College."

than it could get from any fall—not so much, that is, the speed of a body merely dropping toward the sun by its weight, as that of a missile hurled into the orderly solar system from some unknown source without, and also associated with some unknown power; for while it is doubtful whether gravity is sufficient to account for the velocity of all comets, it seems certain that gravity can in no way explain some of the phenomena of their tails.

Thousands of comets have been seen since the Christian era, and the orbits of hundreds have been calculated since the time of Newton. Though they may describe any conic section, and though most orbits are spoken of as parabolas, this is rather a device for the analyst's convenience than the exact representation of fact. Without introducing more technical language, it will be enough to say here that we learn in other cases

FIG. 81.—"A PART OF A COMET."

from the form of the orbit whether the body is drawn essentially by the sun's gravity, or whether it has been thrown into the system by some power beyond the sun's control, to pass away again, out of that control, never to return. It must be admitted, however, that though several orbits are so classed, there is not any one known to be beyond doubt of this latter kind, while we are certain that many comets, if not all, are erratic members of the solar family, coming back again after their excursions, at regular, though perhaps enormous, intervals.

But what we have just been saying belongs rather to the province of the Old Astronomy than the New, which concerns itself more with the nature and appearance of the heavenly bodies than the paths they travel on. Perhaps the best way for us to look at comets will be to confine our attention at first to some single one, and to follow it from its earliest appearance to its last, by the aid of pictures, and thus to study, as it were, the species in the individual. The difficulty will be one which arises from the exquisitely faint and diaphanous appearance of the original, which no ordinary care can possibly render, though here the reader has had done for him all that the wood-engraver can do.

We will take as the subject of our illustration the beautiful comet which those of us who are middle-aged can remember seeing in 1858, and which is called Donati's from the name of its discoverer. We choose this one because it is the subject of an admirable monograph by Professor Bond of the Harvard College Observatory, from which our engravings have, by permission, been made.

Let us take the history of this comet, then, as a general type of others; and to begin at the beginning, we must make the very essential admission that the origin of the comet's life is unknown to us. Where it was born, or how it was launched on its eccentric path, we can only guess, but do not know; and how long it has been traversing it we can only tell later. On the 2d of June, 1858, this one was discovered in the way most comets are found, that is, by a *comet-hunter*, who detected it as a telescopic speck long before it became visible to the naked eye, or put forth the tail which was destined to grow into the beautiful object many of us can remember seeing. For over a century now there has been probably no year in which the heavens have not been thus searched by a class of observers who make comet-hunting a specialty.

FIG. 82. — COMET OF DONATI, SEPT. 24, 1858. (TELESCOPIC VIEW OF HEAD.)

The father of this very valuable class of observers appears to have been Messier, a Frenchman of the last century and of the purest type of the comet-hunters, endowed by Nature with the instinct for their search that a terrier has for rats. In that grave book, Delambre's "History of Astronomy," as we plod along its dry statements and through its long equations, we find, unexpected as a joke in a table of logarithms, the following piece of human nature (quoted from Messier's contemporary, La Harpe) : —

"He [Messier] has passed his life in nosing out the tracks of comets. He is a very worthy man, with the simplicity of a baby. Some years ago he lost his wife, and his attention to her prevented him from discovering a comet he was on the search for, and which Montaigne of Limoges got away from him. He was in despair. When he was condoled with on the loss he had met, he replied, with his head full of the comet, 'Oh, dear! to think that when I had discovered twelve, this Montaigne should have got my thirteenth.' And his eyes filled with tears, till, remembering what it was he ought to be weeping for, he moaned, 'Oh, my poor wife!' but went on crying for his comet."

Messier's scientific posterity has greatly multiplied, and it is rare now for a comet to be seen by the naked eye before it has been caught by the telescope of one of these assiduous searchers. Donati had, as we see, observed his some months before it became generally visible, and accordingly the engraving on page 201 shows it as it appeared on the evening of September 16, 1858, when the tail was already formed, and, though small, was distinct to the naked eye, near the stars of the Great Bear. The reader will easily recognize in the plate the familiar "dipper," as the American child calls it, where the leading stars are put down with care, so that he may, if he please, identify them by comparison with the originals in the sky, even to the little companion to Mizar (the second in the handle of the "dipper," and which the Arabs say is the lost Pleiad). We would suggest that he should note both the length of the tail on this evening as compared with the space between any two stars of the "dipper" (for instance, the two right-hand ones, called the "pointers") and its distance from them, and then turn to page 209, where we have the same comet as seen a little over a fortnight later, on October 3d. Look first at its new place among the stars. The "dipper" is still in view, but the comet has drifted away from it toward the left and into other constellations. The large star close to the left margin of the plate, with three little stars below and to the right, is Arcturus; and the western stars of the Northern Crown are just seen higher up. Fortunately the "pointers," with which we compared the comet on September 16th, are still here, and we can see for ourselves how it has not only shifted but grown. The tail is three times as long as before. It is rimmed with light on its upper edge, and fades away so gradually below that one can hardly say where it ends. But, — wonderful and incomprehensible feature! — shot out from the head, almost as straight as a ray of light itself, but fainter than the moonbeam, now appears an extraordinary addition, a sort of spur, which we can hardly call a new tail, it is so unlike the old one, but which appears to have been darted out into space as if by some mysterious force acting through the head itself. What the spur is, what the tail is, even what the nucleus is, we cannot be said really to know even to-day; but of the tail and of the nucleus or speck in the very head of the comet (too small to be visible in the engraving), we may say that the hairy tail (*comes*) gives the comet its name, and *is* the comet to popular apprehension, but that it is probably the smallest part of the whole mass, while the little shining head, which to the telescope presents a still smaller speck called the nucleus, contains, it now seems probable, the only element of possible danger to the earth.

While admitting our lack of absolute knowledge, we may, if we agree that meteorites were once part of a comet, say that it

FIG. 83. — COMET OF DONATI, OCT. 3, 1858.

now seems probable that the nucleus is a hard, stone-like mass, or collection of such masses, which comes from "space" (that is, from we don't know how far) to the vicinity of the sun, and there is broken by the heat as a stone in a hot fire. (Sir Isaac Newton calculates, in an often quoted passage of the Principia, that the heat which the comet of 1680 was subjected to in its passage by the sun was two thousand times that of red-hot iron.) We have seen the way in which meteoric stones actually do crack in pieces with heat in our own atmosphere, partly, perhaps, from the expansion of the gases the stone contains, and it seems entirely reasonable to suppose that they may do so from the heat of the sun, and that the escaped gases may contribute something toward the formation of the tail, which is always turned away from the sun, and which always grows larger as that is approached, and smaller as it is receded from. However this may be, there is no doubt that the original solid which we here suppose may form the nucleus is capable of mischief, for it is asserted that it often passes the earth's orbit with a velocity of as much as one hundred times that of a cannon-ball; that is, with ten thousand times the destructive capacity of a ball of the same weight shot from a cannon.

One week later, October 9th, the comet had passed over Arcturus with a motion toward our left into a new region of the sky, leaving Arcturus, which we can recognize with the upper one of its three little companions, on the right. Above it is the whole sickle of the Northern Crown, and over these stars the extremity of the now lengthened tail was seen to spread, but with so thin a veil that no art of the engraver can here adequately represent its faintness. The tail then, as seen in the sky, was now nearly twice its former size, though for the reason mentioned it may not appear so in our picture. It should be understood, too, that even the brightest parts of the original were far fainter than they seem here in comparison with the stars, which in the sky are brilliant points of light, which the engraver can only represent by dots of the whiteness of the paper. This being observed, it will be better understood that in the sky itself the faintest stars were viewed apparently undimmed through the brighter parts of the comet, while we can but faintly trace here another most faint but curious feature, a division of the tail into faint cross-bands like auroral streamers, giving a look as if it were yielding to a wind, which folded it into faint ridges like those which may be seen in the smoke of a steamer as it lags far behind the vessel. In fact, when we speak of "the" tail, it must be understood, as M. Faye reminds us, to be in the same sense that we speak of the plume of smoke that accompanies an ocean steamer, without meaning that it is the same thing which we are watching from night to night, more than we do that the same smoke-particles accompany the steamer as it moves across the Atlantic. In both cases the form alone probably remains; the thing itself is being incessantly dissipated and renewed. There is no air here, and yet some of these appearances in the original almost suggest the idea of a medium inappreciably thin as compared with the head of the comet, but whose resistance is seen in the more unsubstantial tail, as that is drawn through it and bent backward, as if by a wind blowing toward the celestial pole.

The most notable feature, however, is the development of a second ray or spur, which has been apparently darted through millions of miles in the interval since we looked at it, and an almost imperceptible bending backward in both, as if they too felt the resistance of something in what we are accustomed to think of as an absolute and perfect void. These tails are peculiarly mysterious features. They are apparently shot out in a direction opposite to the sun (and consequently opposed to the direction of gravity) at the rate of millions of miles a day.

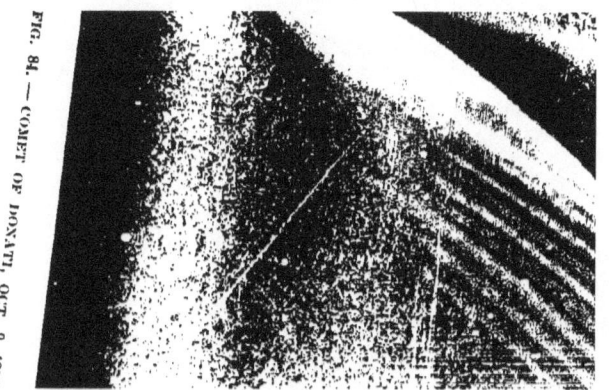

FIG. 84. — COMET OF DONATI, OCT. 9, 1858.

Beyond the fact that the existence of some *repulsive* force in the sun, a "negative gravity" actually existent, not in fancy, but in fact, seems pointed at, astronomers can offer little but conjecture here; and while some conceive this force as of an electrical nature, others strenuously deny it. We ought to admit that up to the present time we really know nothing about it, except that it exists.

At this date (October 9th) the comet had made nearly its closest approach to the earth, and the general outline has been compared to that of the wing of some bird, while the actual size was so vast that even at the distance from which it was seen it filled an angle more than half of that from the zenith to the horizon.

All the preceding drawings have been from naked-eye views; but if the reader would like to look more closely, he can see on page 217 one taken on the night of October 5th through the great telescope at Cambridge, Mass. We will leave this to tell its own story, only remarking that it is not possible to reproduce the phantom-like faintness of the original spur, here also distinctly seen, or indeed to indicate fairly the infinite tenuity of the tail itself. Though millions of miles thick, the faintest star is yet perceptibly undimmed by it, and in estimating the character and quantity of matter it contains, after noting that it is not self-luminous, but shines only like the moon by reflected sunlight, we may recall the acute observation of Sir Isaac Newton where he compares the brightness of a comet's tail with that of the light reflected from the particles in a sunbeam an inch or two thick, in a darkened room, and, after observing that if a little sphere of common air one inch in diameter were rarified to the degree which must obtain at only four thousand miles from the earth's surface it would fill all the regions of the planets to far beyond the orbit of Saturn, suggests the excessively small quantity of vapor that is really requisite to create this prodigious phantom.

The writer has had occasion for many years to make a special study of the reflection of light from the sky; and if such studies may authorize him to express any opinion of his own, he would give his adhesion to the remark of Sir John Herschel, that the actual weight of matter in such a cometary tail may be conceivably only an affair of pounds or even ounces. But if this is true of the tail, it does not follow of the nucleus, just seen in this picture, but of which the engraving on page 205 gives a much more magnified view. It is a sketch of the head alone, taken from a telescopic view on the 24th of September. Here the direction of the comet is still toward the sun (which must be supposed to be some indefinite distance beyond the upper part of the drawing), and we see that the lucid matter appears to be first jetted up, and then forced backward on either side, as if by a wind *from* the sun, to form the tail, presenting successive crescent-shaped envelopes of decreasing brightness, which are not symmetrical, but one-sided, while sometimes the appearance is that of spurts of luminous smoke, wavering as if thrown out of particular parts of the internal nucleus "like a squib not held fast." Down the centre of the tail runs a wonderfully straight black line, like a shadow cast from the nucleus. Only the nucleus itself still evades us, and even in this, the most magnified view which the most powerful telescope till lately in existence could give, remains a point.

Considering the distance of the comet and the other optical conditions, this is still perfectly consistent with the possibility that it may have an actual diameter of a hundred miles or more. It "may" have, observe, not it "has," for in fact we know nothing about it; but that it is at any rate less than some few hundred miles in diameter, and it may, for anything we can positively say, not be more than a very large stone, in which case our atmosphere would probably act as an efficient buffer if it struck us; or it may have a mass which, coupled with its ter-

FIG. 85. — COMET OF DONATI, OCT. 5, 1858. (TELESCOPIC VIEW.)

rible speed, would cause the shock of its contact not so much to pulverize the region it struck, as dissipate it and everything on it instantly into vapor.

Of the remarkable investigations of the spectroscope on comets, we have only room left to say that they inform us that the most prominent cometary element seems to be carbon, — carbon, which Newton two hundred years before the spectroscope, and before the term "carbonic-acid gas" was coined, by some guess or divination had described in other words as possibly brought to us by comets to keep up the carbonic-acid-gas supply in our air, — carbon, which we find in our own bodies, and of which, according to this view, the comets are original sources.

That *we* may be partly made of old and used-up comets, — surely it might seem that a madder fancy never came from the brain of a lunatic at the full of the moon!

Science may easily be pardoned for not giving instant reception to such an idea, but let us also remember, first, that it is a consequence of that of Sir Isaac Newton, and that in the case of such a man as he we should not be hasty to think we understand his ignorance, when we may be "ignorant of his understanding;" and, second, that it has been rendered at least debatable by Dr. Hunt's recent researches whether it is possible to account for the perennial supply of carbon from the earth's atmosphere, without looking to some means of renewal external to the planet.

The old dread of comets is passing away, and all that science has to tell us of them indicates that, though still fruitful sources of curiosity and indeed of wonder, they need no longer be objects of terror. Though there be, as Kepler said, more comets in the sky than fish in the ocean, the encounter of the earth with a comet's tail would be like the encounter with a shadow, and the chance of a collision with the nucleus is remote indeed. We may sleep undisturbed even if a new comet is announced every month, though it is true that here as elsewhere lie remote possibilities of evil.

The consideration of the unfamiliar powers certainly latent in Nature, such as belong to a little tremor of the planet's surface or such as was shown in that scene I have described, when the comparatively insignificant effect of the few tons of dynamite was to make solid buildings unrealities, which vanished away as quickly as magic-lantern pictures from a screen, may help us to understand that the words of the great poet are but the possible expression of a physical fact, and that "the cloud-capped towers, the gorgeous palaces, the solemn temples,"—and we with them,—may indeed conceivably some day vanish as the airy nothings at the touch of Prospero's wand, and without the warning to us of a single instant that the security of our ordinary lives is about to be broken. We concede this, however, in the present case only as an abstract possibility; for the advance of astronomical knowledge is much more likely to show that the kernel of the comet is but of the bigness of some large meteorite, against which our air is an efficient shield, and the chance of evil is in any case most remote,—in any case only such as may come in any hour of our lives from any quarter, not alone from the earthquake or the comet, but from "the pestilence that walketh in darkness;" from the infinitely little below and within us, as well as from the infinite powers of the universe without.

VIII.

THE STARS.

IN the South Kensington Museum there is, as everybody knows, an immense collection of objects, appealing to all tastes and all classes, and we find there at the same time people belonging to the wealthy and cultivated part of society lingering over the Louis Seize cabinets or the old majolica, and the artisan and his wife studying the statements as to the relative economy of baking-powders, or admiring Tippoo Saib's wooden tiger.

There is one shelf, however, which seems to have some attraction common to all social grades, for its contents appear to be of equal interest to the peer and the costermonger. It is the representation of a *man* resolved into his chemical elements, or rather an exhibition of the materials of which the human body is composed. There is a definite amount of water, for instance, in our blood and tissues, and there on the shelf are just so many gallons of water in a large vessel. Another jar shows the exact quantity of carbon in us; smaller bottles contain our iron and our phosphorus in just proportion, while others exhibit still other constituents of the body, and the whole reposes on the shelf as if ready for the coming of a new Frankenstein to re-create the original man and make him walk about again as we do. The little vials that contain the different elements which we all bear about in small proportions are more numerous, and they suggest, not merely the complexity of our constitutions, but the identity of our elements with those we have found by the spectroscope, not alone in the sun, but even in the distant stars and nebulæ; for this wonderful instrument of the New Astronomy can find the traces of poison in a stomach or analyze a star, and its conclusions lead us to think that the ancients were nearly right when they called man a microcosm, or little universe. We have literally within our own bodies

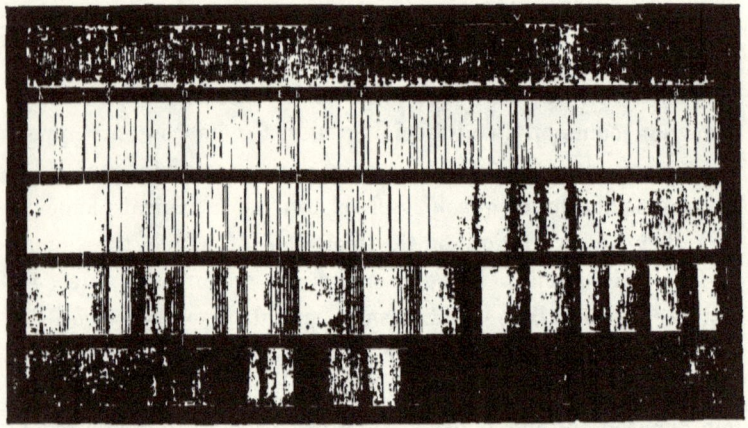

FIG. 86.— TYPES OF STELLAR SPECTRA.

samples of the most important elements of which the great universe without is composed; and you and I are not only like each other, and brothers in humanity, but children of the sun and stars in a more literal sense, having bodies actually made in large part of the same things that make Sirius and Aldebaran. They and we are near relatives.

But if near in kind, we are distant relatives in another way, for the sun, whose remoteness we have elsewhere tried to give an idea of, is comparatively close at hand; quite at hand, one may say, for if his distance, which we have found so enormous, be represented by that of a man standing so close beside us that our hand may rest on his shoulder, to obtain the proportionate distance of one of the *nearest* stars, like Sirius, for instance, we should need to send the man over a hundred miles away. It is probably impossible to give to any one an adequate idea of the extent of the sidereal universe; but it certainly is especially hard for the reader who has just realized with difficulty the actual immensity of the distance of the sun, and who is next told that this distance is literally a physical point as seen from the nearest star. The jaded imagination can be spurred to no higher flight, and the facts and the enormous numbers that convey them will not be comprehended.

Look down at one of the nests of those smallest ants, which are made in our paths. To these little people, we may suppose, the other side of the gravel walk is the other side of the world, and the ant who has been as far as the gate, a greater traveller than a man who comes back from the Indies. It is very hard to think not only of ourselves as relatively far smaller than such insects, but that, less than such an ant-hill is to the whole landscape, is our solar system itself in comparison with the new prospect before us; yet so it is.

All greatness and littleness are relative. When the traveller from the great star Sirius (where, according to the author of "Micromegas," all the inhabitants are proportionately tall and proportionately long-lived), discovered our own little solar system, and lighted on what we call the majestic planet Saturn, he was naturally astonished at the pettiness of everything compared with the world he had left. That the Saturnian inhabitants were in his eyes a race of mere dwarfs (they were only a mile high, instead of twenty-four miles like himself) did not make them contemptible to his philosophic mind, for he reflected that such little creatures might still think and reason; but when he learned that these puny beings were also correspondingly short-lived, and passed but fifteen thousand years between the cradle and the grave, he could not but agree that this was like dying as soon as one was born, that their life was but a span, and their globe an atom. Yet it seems that when one of these very Saturnian dwarfs came afterward with him to our own little ball, and by the aid of a microscope discovered certain animalculæ on its surface, and even held converse with two of them, he could not in turn make up his own mind that intelligence could inhere in such invisible insects, till one of them (it was an astronomer with his sextant) measured his height to an inch, and the other, a divine, expounded to him the theology of some of these mites, according to which all the heavenly host, including Saturn and Sirius itself, were created for *them*.

Do not let us hold this parable as out of place here, for what use is it to write down a long series of figures expressing the magnitude of other worlds, if it leave us with the old sense of the importance to creation of our own; and what use to describe their infinite number to a human mite who reads, and remains of the opinion that *he* is the object they were all created for?

Above us are millions of suns like ours. The Milky Way (shown on page 225) spreads among them, vague and all-surrounding, as a type of the infinities yet unexplored, and of the world of nebulæ of which we still know so little. Let us say at once that it is impossible here to undertake the description of the discoveries of the New Astronomy in this region, for we can scarcely indicate the headings of the chapters which would need to be written to describe what is most important.

The first of these chapters (if we treated our subjects in the order of distance) would be one on space itself, and our changed ideas of the void which separates us from the stars. Of this we will only say in passing, that the old term "the temperature of space" has been nearly abrogated; for while it used to be supposed that more than half of the heat which warmed the earth

FIG. 87. — THE MILKY WAY. (FROM A STUDY BY E. L. TROUVELOT).

came from this mysterious "space" or from the stars, it is now recognized that the earth is principally warmed only by the sun. Of the contents of the region between the earth and the stars, we have, it must be admitted, still little but conjecture; though perhaps that conjecture turns more than formerly to the idea that the void is not a real void, but that it is occupied by something which, if highly attenuated, is none the less matter, and something other and more than the mere metaphysical conception of a vehicle to transmit light to us.

Of the stars themselves, we should need another chapter to tell what has been newly learned as to their color and light, even by the old methods, that is, by the eye and the telescope alone; but if we cannot dwell on this, we must at least refer, however inadequately, to what American astronomers are doing in this department of the New Astronomy, and first in the photometry of the stars, which has assumed a new importance of late years, owing to the labors carried on in this department at Cambridge.

That one star differs from another star in glory we have long heard, but our knowledge of physical things depends largely on our ability to answer the question, "how much?" and the value of this new work lies in the accuracy and fulness of its measures; for in this case the whole heavens visible from Cambridge to near the southern horizon have been surveyed, and the brightness of every naked-eye star repeatedly measured, so that all future changes can be noted. This great work has taxed the resources of a great observatory, and its results are only to be adequately valued by other astronomers; but Professor Pickering's own investigations on variable stars have a more popular interest. It is surely an amazing fact that suns as large or larger than our own should seem to dwindle almost to extinction, and regain their light within a few days or even hours; yet the fact has long been known, while the cause has remained a mystery. A mystery, in most cases, it remains still; but in some we have begun to get knowledge, as in the well-known instance of Algol, the star in the head of Medusa. Here it has always been thought probable that the change was due to something coming between us and the star; but it is on this very account that the new investigation is more interesting, as showing how much can be done on an old subject by fresh reasoning alone, and how much valuable ore may lie in material which has already been sifted. The discussion of the subject by Professor Pickering, apart from its elevated aim, has if, in its acute analysis only, the interest belonging to a story where the reader first sees a number of possible clews to some mystery, and then the gradual setting aside, one by one, of those which are only loose ends, and the recognition of the real ones which lead to the successful solution. The skill of the novelist, however, is more apparent than real, since the riddle he solves for us is one he has himself constructed, while here the enigma is of Nature's propounding; and if the solution alone were given us, the means by which it is reached would indeed seem to be inexplicable.

This is especially so when we remember what a point there is to work on, for the whole system reasoned about, though it may be larger than our own, is at such a distance that it appears, literally and exactly, far smaller to the eye than the point of the finest sewing-needle; and it is a course of accurate reasoning, and reasoning alone, on the character of the observed changing brightness of this point, which has not only shown the existence of some great dark satellite, but indicated its size, its distance from its sun, its time of revolution, the inclination of its orbit, and still more. The existence of dark invisible bodies in space, then, is in one case at least demonstrated, and in this instance the dark body is of enormous size; for, to illustrate by our own solar system, we should probably have to represent it in imagination by a planet or swarm of planetoids hundreds of times the size of Jupiter, and (it may be added) whirling around the sun at less than a tenth the distance of Mercury.

Of a wholly different class of variables are those which have till lately only been known at intervals of centuries, like that new star Tycho saw in 1572. I infer from numerous inquiries that there is such a prevalent popular notion that the "Star of Bethlehem" may be expected to show itself again at about the present time, that perhaps I may be excused for answering these questions in the present connection.

In the first place, the idea is not a new, but a very old one, going back to the time of Tycho himself, who disputed the alleged identity of his star with that which appeared to the shepherds at the Nativity. The evidence relied on is, that bright stars are said to have appeared in this constellation repeatedly at intervals of from three hundred and eight to three hundred and nineteen years (though even this is uncertain); and as the mean of these numbers is about three hundred and fourteen, which again is about one-fifth of 1572 (the then number of years from the birth of Christ), it has been suggested, in support of the old notion, that the Star of Bethlehem might have been a variable, shining out every three hundred and fourteen or three hundred and fifteen years, whose fifth return would fall in with the appearance that Tycho saw, and whose *sixth* return would come in 1886 or 1887. This is all there is about it, and there is nothing like evidence, either that this was the star seen by the Wise Men, or that it is to be seen again by us. On the other hand, nothing in our knowledge, or rather in our ignorance, authorizes us to say positively it cannot come again; and it may be stated for the benefit of those who like to believe in its speedy return, that if it does come, it will make its appearance some night in the northern constellation of Cassiopeia's chair, the position originally determined by Tycho at its last appearance, being twenty-eight degrees and thirteen minutes from the pole, and twenty-six minutes in right ascension.

We were speaking of these new stars as having till lately only appeared at intervals of centuries; but it is not to be inferred that if they now appear oftener it is because there are more of them. The reason is, that there are more persons looking for them; and the fact is recognized that, if we have observers enough and look closely enough, the appearance of "new stars" is not so very rare a phenomenon. Every one at all interested in such matters remembers that in 1866 a new star broke out in the Northern Crown so suddenly that it was shining as bright as the Polar Star, where six hours before there had been nothing visible to the eye. Now all stars are not as large as our sun, though some are much larger; but there are circumstances which make it improbable that this was a small or near object, and it is well remembered how the spectroscope showed the presence of abnormal amounts of incandescent hydrogen, the material which is perhaps the most widely diffused in the universe (and which is plentiful, too, in our own bodies), so that there was some countenance to the popular notion that this was a world in flames. We were, at any rate, witnessing a catastrophe which no earthly experience can give us a notion of, in a field of action so remote that the flash of light which brought the news was unknown years on the way, so that all this — strange but now familiar thought — occurred long before we saw it happen. The star faded in a few days to invisibility to the naked eye, though not to the telescope; and, in fact, all these phenomena at present appear rather to be enormous and sudden enlargements of the light of existing bodies than the creation of absolutely new ones; while of these "new stars" the examples may almost be said to be now growing numerous, two having appeared in the last two years.

Not to enlarge, then, on this chapter of photometry, let us

FIG. 88.—SPECTRA OF STARS IN PLEIADES.

add, in reference to another department of stellar astronomical work, that the recognized master in the study of double stars the world over is not an astronomer by profession, at the head of some national observatory in Berlin or Paris, but a stenographer in the Chicago law-courts, Mr. W. S. Burnham, who, after his day's duties, by nightly labor, prolonged for years with the small means at an amateur's command, has perhaps added more to our knowledge of his special subject in ten years than all other living astronomers.

We have here only alluded to the spectroscope in its application to stellar research, and we cannot now do more than to note the mere headlines of the chapters that should be written on it.

First, there is the memorable fact that, after reaching across the immeasurable distances, we find that the stars are like *us*, — like in their ultimate elements to those found in our own sun, our own earth, our own bodies. Any fuller view of the subject than that which we here only indicate, would begin with the evidence of this truth, which is perhaps on the whole the most momentous our science has brought us, and with which no familiarity should lessen our wonder, or our sense of its deep and permanent significance.

Next, perhaps, we should understand that, invading the province of the Old Astronomy, the spectroscope now tells us of the motions of these stars, which we cannot see move, — motions in what we have always called the "fixed" stars, to signify a state of fixity to the human eye, which is such, that to it at the close of the nineteenth century they remain in the same relative positions that they occupied when that eye first looked on them, in some period long before the count of centuries began.

In perhaps the earliest and most enduring work of man's hands, the great pyramid of Egypt, is a long straight shaft, cut slopingly through the solid stone, and pointing, like a telescope, to the heavens near the pole. If we look through it now we see — nothing; but when it was set up it pointed to a particular star which is no longer there. That pyramid was built when the savages of Britain saw the Southern Cross at night; and the same slow change in the direction of the earth's axis, that in thousands of years has borne that constellation to southern skies, has carried the stone tube away from the star that it once pointed at. The actual motion of the star itself, relatively to our system, is slower yet, — so inconceivably slow that we can hardly realize it by comparison with the duration of the longest periods of human history. The stone tube was pointed at the star by the old Egyptians, but "Egypt itself is now become the land of obliviousness, and doteth. Her ancient civility is gone, and her glory hath vanished as a phantasma. She poreth not upon the heavens, astronomy is dead unto her, and knowledge maketh other cycles. Canopus is afar off, Memnon resoundeth not to the Sun, and Nilus heareth strange voices." In all this lapse of ages, the star's own motion could not have so much as carried it across the mouth of the narrow tube. Yet a motion to or from us of this degree, so slow that the unaided eye could not see it in thousands of years of watching, the spectroscope, first efficiently in the hands of the English astronomer, Dr. Huggins, and later in those of Professor Young of Princeton, not only reveals at a look, but tells us the amount and direction of it, in a way that is as strange and unexpected, in the view of our knowledge a generation ago, as its revelation of the essential composition of the bodies themselves.

Again, in showing us this composition, it has also shown us more, for it has enabled us to form a conjecture as to the relative ages of the stars and suns; and this work of classifying them, not only according to their brightness, but each after his kind, we may observe was begun by a countryman of our own, Mr. Rutherfurd, who seems to have been among the first after

THE STARS.

Fraunhofer to apply the newly-invented instrument to the stars, and quite the first to recognize that these were, broadly speaking, divisible into a few leading types, depending not on their size but on their essential nature. After him Secchi (to whom the first conception is often wrongly attributed) developed it, and gave four main classes into which the stars are in this way divisible, a classification which has been much extended by others; while the first carefully delineated spectra were those of Dr. Huggins, who has done so much for all departments of our science that in a fuller account his name would reappear in every chapter of this New Astronomy, and than whom there is no more eminent living example of its study. Owing to their feeble light, years

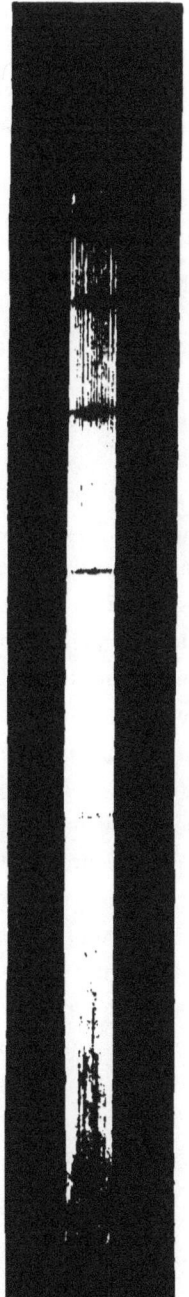

FIG. 29.—SPECTRUM OF ALDEBARAN.

FIG. 30.—SPECTRUM OF VEGA.

were needed when he began his work to depict completely so full a single spectrum as that he gives of Aldebaran, though he has lived to see stellar spectrum photography, whose use he first made familiar, producing in its newest development, which we give here, the same result in almost as many minutes. Before we present this latest achievement of celestial photography, let us employ the old method of an engraving made from eye-drawings, once more, to illustrate on page 222 the distinct character of these spectra, and their meaning. In the telespectroscope, the star is drawn out into a band of colored light, but here we note only in black and white the lines which are seen crossing it, the red end in these drawings being at the left, and the violet at the right; and we may observe of this illustration, that though it may be criticised by the professional student, and though it lack to the general reader the attraction of color, or of beautiful form, it is yet full of interest to any one who wishes to learn the meaning of the message the star's light can be made to yield through the spectroscope, and to know how significant the differences are it indicates between one star and another, where all look so alike to the eye. First is the spectrum of a typical white or blue-white star, Sirius, — the very brightest star in the sky, and which we all know. The brighter part of the spectrum is a nearly continuous ribbon of color, crossed by conspicuous, broad, dark lines, exactly corresponding in place to narrower ones in our sun, and due principally to hydrogen. Iron and magnesium are also indicated in this class, but by too fine lines to be here shown.

Sirius, as will be presently seen, belongs to the division of stars whose spectrum indicates a very high temperature, and in this case, as in what follows, we may remark (to use in part Mr. Lockyer's words) that one of the most important distinctions between the stars in the heavens is one not depending upon their mass or upon anything of that kind, but upon conditions which make their spectra differ, just in the way that in our laboratories the spectrum of one and the same body will differ at different temperatures.

What these absolutely are in the case of the stars, we may not know; but placing them in their most probable relative order, we have taken as an instance of the second class, or lower-temperature stage, our own sun. The impossibility of giving a just notion of its real complexity may be understood, when we state that in the recent magnificent photographs by Professor Rowland, a part alone of this spectrum occupies something like fifty times the space here given to the whole, so that, crowded with lines as this appears, scarcely one in fifty of those actually visible can be given in it. Without trying to understand all these now, let us notice only the identity of two or three of its principal elements with those found in other stars, as shown by the corresponding identity of some leading lines. Thus, C and F (with others) are known to be caused by hydrogen; D, by sodium; b, by magnesium; while fainter lines are given by iron and by other substances. These elements can be traced by their lines in most of the different star-spectra on this plate, and all those named are constituents of our own frames.

The hydrogen lines are not quite accurately shown in the plate from which our engraving is made, those in Sirius, for instance, being really wider by comparison than they are here given; and we may observe in this connection, that by the particular appearance such lines wear in the spectrum itself we can obtain some notion of the *mass* of a star, as well as of its chemical constitution. We can compare the essential characteristics of such bodies, then, without reference to their apparent size, or as though they were all equally remote; and it is a striking thought, that when we thus rise to an impartial contemplation of the whole stellar universe, our sun, whose least ray makes the whole host of stars disappear, is found to be not only itself a star, but by comparison a small one,— one at least which is more probably below than above the average individual of its class, while some, such as Sirius, are not impossibly hundreds of times its size.

Then comes a third class, such as is shown in the spectrum of the brightest star in Orion, looking still a little like that of our sun; but yet more distinctively in that of the brightest star in Hercules, looking like a columnar or fluted structure, and concerning which the observations of Lockyer and others create the strong presumption, not to say certainty, that we have here a lower temperature still. Antares and other reddish stars belong to this division, which in the very red stars passes into the fourth type, and there are more classes and subclasses without end; but we invite here attention particularly to the first three, much as we might present a child, an adult, and an old man, as types of the stages of human existence, without meaning to deny that there are any number of ages between. We can even say that this may be something more than a mere figure of speech, and that a succession in age is not improbably pointed at in these types.

We may have considered — perhaps not without a sort of awe at the vastness of the retrospect — the past life of the worlds of our own system, from our own globe of fluid fire as we see it by analogy in the past, through the stages of planetary life to the actual condition of our present green earth, and on to the stillness of the moon. Yet the life history of our sun, we can hardly but admit, is indefinitely longer than this. We feel, rather than comprehend, the vastness of the period that separates our civilization from the early life of the world; but what is this to the age of the sun, which has looked on and seen its planetary children grow? Yet if we admit this temperature classification of the stars, we are not far from admitting that the

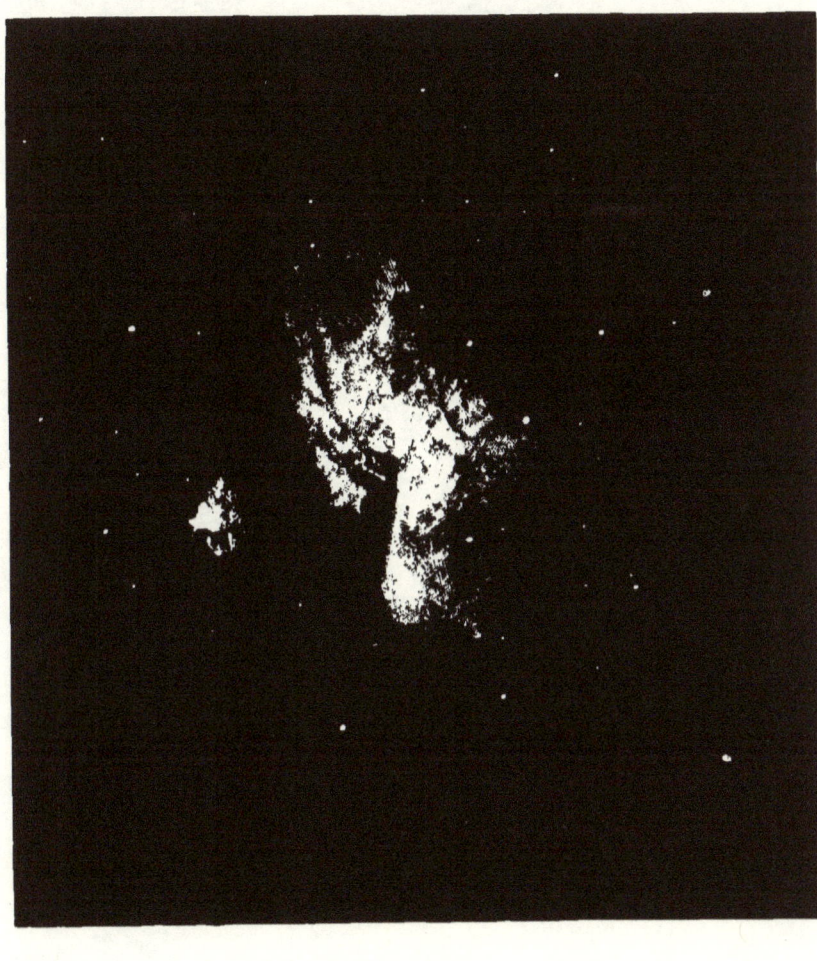

FIG. 91.—GREAT NEBULA IN ORION. (FROM A PHOTOGRAPH BY A. A. COMMON, F.R.S.)

spectroscope is now pointing out the stages in the life of suns themselves; suns just beginning their life of almost infinite years; suns in the middle of their course; suns which are growing old and casting feebler beams, — all these and many more it brings before us.

Another division of our subject would, with more space, include a fuller account of that strange and most interesting development of photography which is going on even while we write; and this is so new and so important, that we must try to give some hint of it even in this brief summary, for even since the first chapters of this book were written, great advances have taken place in its application to celestial objects.

Most of us have vague ideas about small portions of time; so much so, that it is rather surprising to find to how many intelligent people, a second, as seen on the clock face, is its least conceivable interval. Yet a second has not only a beginning, middle, and end, as much as a year has, but can, in thought at least, be divided into just as many numbered parts as a year can. Without entering on a disquisition about this, let us try to show by some familiar thing that we can at any rate not only divide a second in imagination into, let us say, a hundred parts, but that we can observe distinctly what is happening in such a short time, and make a picture of it, — a picture which shall be begun and completed while this hundredth of a second lasts.

Every one has fallen through at least some such a little distance as comes in jumping from a chair to the floor, and most of us, it is safe to say, have a familiar impression of the fact that it takes, at any rate, less than a second in such a case from the time the foot leaves its first support till it touches the ground. Plainly, however large or small the fall may be, each fraction of an inch of it must be passed through in succession, and if we suppose the space to be divided, for instance, into a hundred parts, we must divide in thought the second into at least as many, since each little successive space was traversed in its own little interval of time, and the whole together did not make a second. We can even, as a matter of fact, very easily calculate the time that it will take anything which has already fallen, let us say one foot, to fall an inch more; and we find this, in the supposed instance, to be almost exactly one one-hundredth of a second. On page 243 is a reproduction of a photograph from Nature, of a man falling freely through the air. He has dropped from the grasp of the man above him, and has already fallen through some small distance, — a foot or so. If we suppose it to be a foot, since we can see that the man's features are not blurred, as they would undoubtedly have been had he moved even much less than an inch while this picture was being taken, it follows, from what has been said, that the making of the whole picture — landscape, spectators, and all — occupied not *over* one one-hundredth of a second.

We have given this view of "the falling man" because, rightly understood, it thus carries internal evidence of the limit of time in which it could have been made; and this will serve as an introduction to another picture, where probably no one will dispute that the time was still shorter, but where we cannot give the same kind of evidence of the fact.

"Quick as lightning" is our common simile for anything occupying, to ordinary sense, no time at all. Exact measurements show that the electric spark does occupy a time, which is almost inconceivably small, and of which we can only say here that the one one-hundredth of a second we have just been considering is a long period by comparison with the duration of the brightest portion of the light.

On page 245 we have the photograph of a flash of lightning (which proves to be several simultaneous flashes), taken last July from a point on the Connecticut coast, and showing not

FIG. 92. — A FALLING MAN.

only the vivid zigzag streaks of the lightning itself, but something of the distant sea view, and the masts of the coast survey schooner "Palinurus" in the foreground, relieved against the sky. We are here concerned with this interesting autograph of the lightning, only as an illustration of our subject, and as proving the almost infinite sensitiveness of the recent photographic processes; for there seems to be no limit to the briefness of time in which these can so act in some degree, whether the light be bright or faint, and no known limit to the briefness of time required for them to act *effectively* if the light be bright enough.

What has just preceded will now help us to understand how it is that photography also succeeds so well in the incomparably fainter objects we are about to consider, and which have been produced not by short but by long exposures. We have just seen how sensitive the modern plate is, and we are next to notice a new and very important point in which photographic action in general differs remarkably from that of the eye. Seeing may be described, not wholly inaptly, as the recognition of a series of brief successive photographs, taken by the optic lens on the retina; but the important difference between seeing and photographing, which we now ask attention to, is this: When the eye looks at a faint object, such as the spectrum of a star, or at the still fainter nebula, this, as we know, appears no brighter at the end of half an hour than at the end of the first half-second. In other words, after a brief fraction of a second, the visual effect does not sensibly accumulate. But in the action of the photograph, on the contrary, the effect *does* accumulate, and in the case of a weak light accumulates indefinitely. It is owing to this precious property, that supposing (for illustration merely) the lightning flash to have occupied the one-thousandth part of a second in impressing itself on the plate, to get a nearly similar effect from a continuous light one thousand times weaker, we have only to expose the plate a thousand times as long, that is, for one second; while from a light a million times weaker we should get the same result by exposing it a million times as long, that is, for a thousand seconds.

And now that we come to the stars, whose spectra occupy minutes in taking, what we just considered will help us to understand how we can advantageously thus pass from a thousandth of a second or less, to one thousand seconds or even more, and how we can even, — given time enough, —

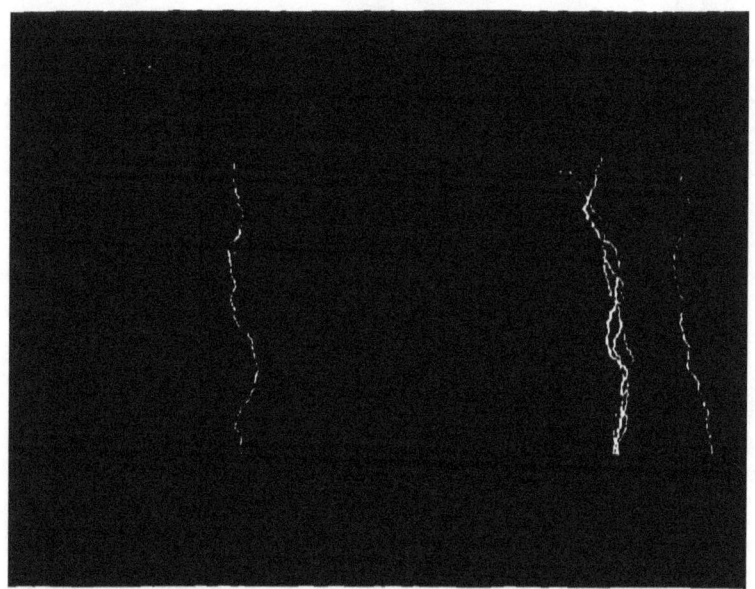

FIG. 93. — A FLASH OF LIGHTNING. (FROM A PHOTOGRAPH BY DR. H. G. PIFFARD.)

conceivably, be able to photograph what the eye *cannot see at all*.

We have on page 231 a photograph quite recently taken at Cambridge from a group of stars (the Pleiades) passing by the telescope. Every star is caught as it goes, and presented, not in its ordinary appearance to the eye, but by its spectrum. There is a general resemblance in these spectra from the same cluster; while in other cases the spectra are of all types and kinds, the essential distinction between individuals alike to the eye, being more strikingly shown, as stars apparently far away from one another are seen to have a common nature, and stars looking close together (but which may be merely in line, and really far apart) have often no resemblance; and so the whole procession passes through the field of view, each individual leaving its own description. This self-description will be better seen in the remarkable photographs of the spectra of Vega and Aldebaran, which are reproduced on page 235 from the originals by a process independent of the graver. They were obtained on the night of November 9, 1886, at Cambridge, as a part of the work pursued by Professor Pickering, with means which have been given from fitting hands, thus to form a memorial of the late Dr. Henry Draper. We are obliged to the source indicated, then, for the ability to show the reader here the latest, and as yet inedited, results in this direction; and they are such as fully to justify the remark made above, that minutes, by this new process, take the place of years of work by the most skilful astronomer's eye and hand.

The spectrum of Vega (Alpha Lyræ) is marked only by a few strong lines, due chiefly to hydrogen, because these are all there are to be seen in a star of its class. Aldebaran (the bright star in Taurus), on the contrary, here announces itself as belonging to the family of our own sun, a probably later type, and distinguished by solar-like lines in its spectrum, which may be counted in the original photograph to the number of over two hundred. There is necessarily some loss in the printed reproduction; but is it not a wonderful thing, to be able to look up, as the reader may do, to Aldebaran in the sky, and then down upon the page before us, knowing that that remote, trembling speck of light has by one of the latest developments of the New Astronomy been made, without the intervention of the graver's hand, to write its own autograph record on the page before him?

In the department of nebular astronomy, photography has worked an equal change. The writer well remembers the weeks he has himself spent in drawing or attempting to draw nebulæ, — things often so ghost-like as to disappear from view every time the eye turned from the white paper, and only to be seen again when it had recovered its sensitiveness by gazing into the darkness. The labors of weeks were, literally, only represented by what looked like a stain on the paper; and no two observers, however careful, could be sure that the change between two drawings of a nebula at different dates was due to an alteration in the thing itself, or in the eye or hand of the observer, though unfortunately for the same reason it is impossible fully to render the nebulous effect of the photograph in engraving. We cannot with our best efforts, then, do full justice to the admirable one of Orion, on page 239, which we owe to the particular kindness of Mr. Common, of Ealing, England, whose work in this field is as yet unequalled. The original enlargement measures nearly two square feet in area, with fine definition. It is taken by thirty-nine minutes' exposure, and its character can only be indicated here; for it is not too much to say here of this original also, that as many years of the life of the most skilled artist could not produce so trustworthy a record of this wonder.

The writer remembers the interest with which he heard Dr. Draper, not long before his lamented death, speak of the almost incredible sensitiveness of these most recent photographic processes, and his belief that we were fast approaching the time when we should photograph what we could not even see. That time has now arrived. At Cambridge, in Massachusetts, and at the Paris Observatory, by taking advantage of the cumulative action we have referred to, and by long exposures, photographs have recently been taken showing stars absolutely invisible to the telescope, and enabling us to discover faint nebulæ whose previous existence had not been suspected; and when we consider that an hour's exposure of a plate, now not only secures a fuller star-chart than years of an astronomer's labor, but a more exact one, that the art is every month advancing perceptibly over the last, and that it is already, as we may say, not only making pictures of what we see, but of what we cannot see even with the telescope, — we have before us a prospect whose possibilities no further words are needed to suggest.

We have now, not described, but only mentioned, some division of the labors of the New Astronomy in its photometric, spectroscopic, and photographic stellar researches, on each of which as many books, rather than chapters, might be written, to give only what is novel and of current interest. But these are themselves but a part of the modern work that has overturned or modified almost every conception about the stellar universe which was familiar to the last generation, or which perhaps we were taught in our own youth.

In considering the results to be drawn from this glance we have taken at some facts of modern observation, if it be asked, not only what the facts are, but what lessons the facts themselves have to teach, there is more than one answer, for the moral of a story depends on the one who draws it, and we may look on our story of the heavens from the point of view either of our own importance or of our own insignificance. In the one case we behold the universe as a sort of reflex of our own selves, mirroring in vast proportions of time and space our own destiny; and even from this standpoint, one of the lessons of our subject is surely that there is no permanence in any created thing. When primitive man learned that with lapsing years the oak withered and the very rock decayed, more slowly but as surely as himself, he looked up to the stars as the types of contrast to the change he shared, and fondly deemed them eternal; but now we have found change there, and that probably the star clusters and the nebulæ, even if clouds of suns and worlds, are fixed only by comparison with our own brief years, and, tried by the terms of their own long existence, are fleeting like ourselves.

"We have often witnessed the formation of a cloud in a serene sky. A hazy point barely perceptible — a little wreath of mist increases in volume and becomes darker and denser, until it obscures a large portion of the heavens. It throws itself into fantastic shapes, it gathers a glory from the sun, is borne onward by the wind, and as it gradually came, so, perhaps, it gradually disappears, melting away in the untroubled air. But the universe is nothing more than such a cloud, — a cloud of suns and worlds. Supremely grand though it may seem to us, to the infinite and eternal intellect it is no more than a fleeting mist. If there be a succession of worlds in infinite space, there is also a succession of worlds in infinite time. As one after another cloud replaces clouds in the skies, so this starry system, the universe, is the successor of countless others that have preceded it, — the predecessor of countless others that will follow."

These impressions are strengthened rather than weakened when we come back from the outer universe to our own little solar system; for every process which we know, tends to the dissipation, or rather the degradation, of heat, and seems to point, in our present knowledge, to the final decay and extinction of the light of the world. In the words of one of the most eminent living students of our subject, "The candle of the sun is burning down, and, as far as we can see, must at last reach the socket. Then will begin a total eclipse which will have no end.
'Dies iræ, dies illa,
Solvet sæclum in favilla.'"

Yet though it may well be that the fact itself here is true, it is possible that we draw the moral to it, unawares, from an unacknowledged satisfaction in the idea of the vastness of the funeral pyre provided for such beings as ourselves, and that it is pride, after all, which suggests the thought that when the sun of the human race sets, the universe will be left tenantless, as a body from which the soul has fled. Can we not bring ourselves to admit that there may be something higher than man and more enduring than frail humanity, in some sphere in which *our* universe, conditioned as it is in space and time, is itself embraced; and so distrust the conclusions of man's reason where they seem to flatter his pride?

May we not receive even the teachings of science, as to the "Laws of Nature," with the constant memory that all we know, even from science itself, depends on our very limited sensations, our very limited experience, and our still more limited power of conceiving anything for which this experience has not prepared us?

I have read somewhere a story about a race of ephemeral insects who live but an hour. To those who are born in the early morning the sunrise is the time of youth. They die of old age while its beams are yet gathering force, and only their descendants live on to midday; while it is another race which sees the sun decline, from that which saw its rise. Imagine the sun about to set, and the whole nation of mites gathered under the shadow of some mushroom (to them ancient as the sun itself) to hear what their wisest philosopher has to say of the gloomy prospect. If I remember aright, he first told them that, incredible as it might seem, there was not only a time in the world's youth when the mushroom itself was young, but that the sun in those early ages was in the eastern, not in the western, sky. Since then, he explained, the eyes of scientific ephemera had followed it, and established by induction from vast experience the great "Law of Nature," that it moved only westward; and he showed that since it was now nearing the western horizon, science herself pointed to the conclusion that it was about to disappear forever, together with the great race of ephemera for whom it was created.

What his hearers thought of this discourse I do not remember, but I have heard that the sun rose again the next morning.

INDEX.

ABBE, PROFESSOR, 56.
Actinism, 71.
Adams, Professor, 195.
Africa, 116.
Ages, stellar, 238.
Air: dancing, 17; a medium, 33; continuous, 176; rarefied, 179; motes, 181; nimble, 191. (See *Atmosphere*.)
Airless Mountains, 160.
Air-wave, 185.
Aitken's Researches, 181.
Alaska, 38.
Aldebaran, 222, 235, 236, 246.
Algot, 228.
Allegheny Observatory, 17, 19, 84, 86. (See *Langley*.)
Alphonsus Ring-plain, 156.
Alps, 39, 148, 151, 167, 181. (See *Apennines, Lunar*.)
American Astronomers, 227.
American Continents, 20, 21, 31. (See *South*.)
Andalusia, 53.
Animalculæ, 224.
Animals: food, 74; fright, 42. (See *Dog*.)
Antares, 238.
Ants, 223. (See *Insects*.)
Apennines, 151, 153, 155, 160, 167. (See *Alps, Lunar*.)
Apples, 171.
Arab Traditions, 194. (See *Moslem*.)
Arago, quoted, 41, 42.
Archimedes, 94.
Archimedes Crater, 151–153, 155.
Arctic Cold, 159.
Arctic Pole, 96.
Arcturus, 208, 211.

Aristillus Crater, 151.
Aristotelian Philosophy, 8.
Arzachel, 156, 161.
Asteroids, 128.
Astrology, 127.
Astronomers and Priests, 1–3. (See *American, New, Old*.)
Astronomical Day, 85, 86.
Atmosphere, 136, 180; as a shield, 216, 220. (See *Air*.)
Atolls, 152.
Auger, simile, 31.
Aurora Borealis, 35, 67, 212.
Autolycus Crater, 151.
Axis, 9, 10.

BABEL, 96.
Bain Telegraph, 88.
Balloons, 176.
Bees, 124. (See *Insects*.)
Berkeley's Theory, 70.
Berlin Observatory, 233.
Bernières's Lens, 103.
Bessemer Steel, 104–108.
Birds, 172, 196, 197. (See *Animals*.)
Black Hole, 73.
Bond, Professor, 204.
Boston, Mass., 88, 132.
Bothkamp, observations at, 66.
Breadstuffs, 78, 79. (See *Grain, Sun-spots, Wheat*.)
Bridges, 20, 68.
Britain, Ancient, 1, 234. (See *England*.)
British Isles, 14, 25.
Brocken Spectre, 55.

INDEX.

Brothers, Mr., 50.
Bubbles, 168.
Buffer, the air as a, 216, 220.
Bunsen's Researches, 12.
Burnham, W. S., 233.
Burning-glasses, 102-104.
Burning Heat, 160, 163.

CACTUS, 14, 24.
Calcutta, 73.
California, 151, 180.
Cambric Needle (*q. v.*), experiment, 132.
Cambridge Observations, 227, 245-247.
Camera Obscura, 63.
Campanus Crater, 163, 165.
Candle, simile, 39.
Cannon-ball, 5, 38, 41, 98, 135, 186, 211.
Canopus, 234.
Carbon, 72, 73, 107, 221.
Carbonic-acid Gas, 219.
Carpenter's Studio, 140.
Carrington's Work, 79, 87.
Carthage, 116.
Cassini, 42.
Cassiopeia, 229.
Cataclysm, 30.
Centimetres, 93.
Chacornac's Drawing, 33.
Chambers, on sun-spots, 80.
Charleston Earthquake (*q. v.*), 42.
Chemical Elements, 221, 223.
Cherry-stone, comparison, 196.
Chicago: great fire, 134; astronomer, 233.
China: lens, 103, 104; soil, 180.
Chlorophyl, 73.
Chocolate, simile, 107.
Cholera, 80.
Chromosphere, 7; clouds, 62; forms, 64-68.
Cinders, 171.
Clark's Glasses, 123.
Cliffs, 164.
Clock, 135.
Cloud-ocean, 179.
Clouds: cirrous, 27, 28; beautiful, 54; and rain, 111; formed, 249.
Coal-beds, 115.
Coal: energy, 73-75, 111; destroyed, 97; wasted, 101; stock, 112.
Cobweb, simile, 26.
Cold: and eclipses, 40; in planets, 136.

Colorado, 50.
Colors: in eclipses (*q. v.*), 65; mental, 70, 71; in Jupiter (*q. v.*), 127; in moon (*q. v.*), 168; in stars (*q. v.*), 227; spectrum (*q. v.*), 236.
Comet-hunters, 204, 207.
Comets: chapter, 199-220; Donati's, 201, 204, 205, 207, 209, 217; one part, 203; parts and name, 208; tail (*q. v.*), 208, 211; diameter and parts, 216; spectroscope, elements, dread, 219; numerous, stone, 219, 220; kernel, 220; (1858), 213-216; (1866), 200.
Common, A. A., 239, 247.
Compass, 86.
Connecticut Observations, 186, 242.
Converter, 104-108.
Coral, 151.
Corn, 111. (See *Grain*.)
Corona, 7, 36, 37, 40, 41, 43, 45-52, 55, 56, 59, 60-62.
Cotton-mill, 74.
Counting, 94.
Cracks, celestial, 163.
Craters, 164. (See special names.)
Crystalline Structure, 4, 23-27.
Cyclones, 24, 31, 32, 68.

DECAY, 248, 249.
Delambre's History, 207.
De la Rue's Engraving, 125.
Delfthaven, 5.
Denning's Theory, 197.
Diamonds, melted, 103.
Dies Iræ, 249.
Dipper, 207, 208. (See *Great Bear*, *Polar*.)
Diurnal Oscillation, 87.
Dog, anecdote of, 42. (See *Animals*.)
Donati, 201, 204, 205, 207, 209, 213, 217. (See *Comets*.)
Double Stars, 233.
Draper, Professor Henry, 128, 246, 247.
Ducks, noise, 188.
Dust, 34, 100, 101, 102, 105, 197.
Dynamite, 182, 185, 220.

EARTH: relations, 3, 4; description difficult, 6; temperature (*q. v.*), 34, 101; a string of earths, 96; stars like, 118; seen from outside, 133-135; craters, 148.

INDEX. 255

Earthquakes, 220. (See *Charleston.*)
Earth-shine, 167, 172.
Eclipses: total, 7, 37; screen, 36; three, 39, 55; partial, 40; singular gloom, 39–43; causing fright, 43; colors (*q. v.*), 48, 56, 61, 65, 66; (1842), 41; (1857), 48; (1869), 39, 40; (1870), 44, 61; (1871), 50, 66, 68; (1878), 38, 50, 57, 58. (See *Total.*)
Egypt, 116, 234. (See *Pyramids.*)
Electricity, 13, 75, 76.
Electric Light, 7.
Electric Spark, 242. (See *Lightning.*)
Electric Storm, 84, 85, 88.
Elizabeth, Queen, 115.
Engine-power, 98, 111.
England: fleets, 2; coal, 115. (See *Britain, London.*)
Engraving, 17.
Enigma, 228.
Ephemera, 250, 251.
Equatorial Landscape, 13, 17, 18, 47.
Equatorial Telescope, 122.
Ericsson: engravings, 112, 113; discoveries, 163.
Eruptive Promontories, 66–68.
Etna, 164, 181.
Europe, size, 25.
Evolution, planetary, 139.
Explosive Forces, 182–194.
Eye, 71, 227.
Eye-pieces, 47, 63.

FABRICIUS'S OBSERVATIONS, 8.
Fact and Fancy, 175.
Factory, 73.
Faculæ, 32, 33.
Falling, 242, 243.
Falling Stars, 193. (See *Meteors, Shooting.*)
Faraday, Michael, 76.
Fault, technical term, 156.
Faust, 139.
Faye: theory, 29–32; on Comets' Tails, 212.
Fern-like Forms, 25, 26.
Filaments, 25–27, 30, 55, 56, 65, 66, 68.
Fire, in sun (*q. v.*), 92. (See *Flames, Heat.*)
Fixed Stars, 233.
Flame-like Appearances, 23, 24.
Flames, 65, 66, 69, 185.
Flashes, 189, 195.
Flax, 111.

Flowers, color (*q. v.*), 70. (See *Rose, Plants.*)
Foliage-forms, 32.
Fontenelle's Story, 133.
Forbes's Observations, 38, 39.
Frankenstein, 221.
Franklin's Discoveries, 76.
Fraunhofer Studies, 235.
French Institute, 186.
Frost-crystals (*q. v.*), 23.
Furnaces, 101.

GALILEO, 8, 121–123, 139, 140.
Gas: glowing, 44; in sun, 60.
Gas-jets, 40, 61, 68, 88.
Gassendi's View, 172, 173.
Gelinck's Observations, 80.
Geminids, 196.
Genii, 193.
Geographers and Geologists, 133.
Glare, 14, 18, 62–64.
Glass: spun, 26; globe, 145.
Glow-worms, 7, 117.
Good Hope Observations, 80.
Gould's Researches, 80.
Grain, prices, 77, 80, 87. (See *Corn, Sun-spots, Wheat.*)
Gramarye, 92.
Grass-blades, 66, 72.
Grasses, 26.
Gravitation, 72, 203; negative, 215.
Great Bear, 207. (See *Dipper, Polar.*)
Green's Maps, 130.
Greenwich Observatory, 2, 81, 82, 84, 85, 88, 89.
Gulliver's Travels, 131, 132. (See *Swift.*)
Gunpowder, 186.
Guns, 135. (See *Cannon-ball.*)

HALL ISLAND, 130.
Hall, Professor, 131.
Hand, illustration, 168.
Harkness's Observations, 44.
Harvests, 90.
Hastings, Professor, 60.
Heat: development, 13; concentration, 19; loss, 29; confinement, 33; sensation, 71; vibrations, 72; energy, 91; amount, 92, 97; computation, 94–96; diminution, 101; emission, 102; storage, 111; in sugar, 188. (See *Flames, Sun.*)

Hecla, 104, 181.
Hedgehog-spines, simile, 68.
Helmholtz's Estimates, 98.
Hengist and Horsa, 1. (See *Britain*.)
Hercules, 238.
Herschel, Sir John: sun-spots, 12-14; electric storms, 88; comet's tail, 216.
Herschel, Sir William: avoidance of light, 18; prices, 79; sun-spots (*q. v.*), 129.
Herschel's Outlines, 11.
Holden, Professor, 124.
Honeycomb Structure, 30.
Huggins's Experiment, 234, 235.
Humanity, deified, 172.
Human Race, 250.
Humboldt, 195.
Humming-bird, 70.
Hunt, Professor, 136, 219.
Hydrogen, 68, 99, 237.

Ibrahim, King, story, 194, 195.
Ice: melted, 95, 96; never melted, 163, 164.
Imbrian Sea, 151.
Insects, 224, 250. (See *Ants*, *Bees*.)
Iron: melting, 19, 107; appearance of cold, 25; in sun, 28; in man, 221; in stars, 236, 237. (See *Steel*.)
Ironstone, 188.
Ivy, 115.

Janssen's Observations, 61.
Jevons, Professor, 80.
Joseph in Egypt, 90.
Jumping, 241, 242.
Jupiter, 79, 118, 124, 127-129, 156, 185, 229.

Kensington Museum, 221.
Kepler, on Comets, 219.
Kernels, 220.
Kew, 88.
Kirchoff's Researches, 12.
Krakatao, 181, 185, 186.

La Harpe, quoted, 207.
Landscape, 169.
Langley, Prof. S. P.: drawings, 15, 16, 18, 19, 21, 22, 25, 28, 30; note-book, 24; expedition, 180; study of Reflection, 216. (See *Allegheny*, *Pittsburg*.)
Latent Power, 220.
Laws of Nature, 250, 251.
Leaf-like Appearances, 23. (See *Willow*.)
Lenses, 102, 103; Galileo's, 123.
Leo, 195, 197.
Linis's Drawing, 48, 50.
Lick Glass, 123.
Light: development, 13; day and night, 35; white (*q. v.*), 48; mental (see *Eye*), 71; from balloon, 179; transmitted, 227. (See *Sun*.)
Lightning, 75, 76, 242, 244, 245. (See *Electric*.)
Lily, 73. (See *Flowers*.)
Limited Express Train, 5.
Loaf-sugar, experiment, 188.
Lockyer's Land, 130.
Lockyer's Solar Physics, 59, 61, 236, 238.
Lombardy, 151.
London, 111.
Lost Pleiad (*q. v.*), 207.
Louis XV., 42.
Louis XVI., 221.
Lunar Alps (*q. v.*), 148, 149. (See *Moon*.)
Lunar Apennines (*q. v.*), 153.
Lunar Shadows, 36, 37, 39, 56.
Lyrids, 196, 200.

Macartney's Lens, 103.
Maelstrom, 27.
Magic Lantern, simile, 220.
Magnesium, 236, 237.
Magnetic Needle, 81, 82, 84, 85, 87, 89.
Mammoth Cave, 40.
Man, chemistry of, 221, 233. (See *Humanity*.)
Manhattan Island, 111.
Mare Crisium, 143.
Mare Serenitatis, 143, 144.
Mars, 118, 128-132, 148.
Mason's Publication, 137.
Matterhorn, 148, 167.
Mayflower, 5.
Meadows, 172.
Mecca, 175.
Medusa, 228.
Memnon, 234.
Mercator, 163, 165.

INDEX.

Mercury, 3, 118, 136, 229.
Messier, anecdote, 207.
Metals, melted, 103. (See *Iron*.)
Metaphysics, 70, 71.
Meteorites: around Saturn, 124; recent, 187; lawsuit, 187, 188; analyzed, 191, 192; in Iowa, 199, 200; swarm, 200; cracking, 211.
Meteors, 98, 175-198; (1868), 189. (See *Falling, Shooting*.)
Meunier's Investigations, 192.
Mexican Gulf, 38.
Microcosm, 222.
Micromegas, 223.
Microscope, 224.
Middle Ages, 91, 175.
Milky Way, 224, 225.
Milton, quoted, 14, 38.
Mind-causation, 70, 71.
Mirror, 102, 107.
Mississippi, 134.
Mites, 224.
Mizar, 207.
M'Leod's Drawing, 44.
Monochromatic Light (*q. v.*), 63.
Montaigne of Limoges, 207.
Mont Blanc, 156.
Monte Rosa, 167.
Moon: practical observations, 2; newly studied, 3; distance, 4-6; size, 5, 6, 140, 156; shadows (*q. v.*), 36, 125; in sun-eclipse, 41; planetary relations, 117-174; and Jupiter, 127; photograph, 137; full, 141, 144, 147; Man in the, 143; mountains, 144; craters, 147, 148; temperature, 159; airless, 160; landscape (*q. v.*), 169; age, 171; broken up, 192; like comet, 215. (See *Lunar*.)
Moslem Traditions, 175, 194. (See *Arab*.)
Moss, 160.
Mouchot's Engravings, 109, 112.
Mountain Sickness, 50, 53.

NAPLES, 155, 157. (See *Vesuvius*.)
Napoleon, 80, 134.
Nasmyth's Researches, 11, 12, 14, 24, 25, 140.
Nativity of Jesus, 229.
Nature's Laws (*q. v.*), 176.
Nebulæ, 247.
Needle, 228. (See *Cambric*.)
Neptune, 121.

Nerves, none in camera, 47.
Nerve Transmission, 5, 6.
New Astronomy, 4, 75, 76, 117, 121, 171, 193, 222, 224, 227, 235, 246, 248. (See *Old*.)
Newcomb, Professor, 55.
Newspapers, printed by the sun, 74.
Newton, Professor, 191, 195-197.
Newton, Sir Isaac, 136, 203, 211; on Comets, 215, 219.
Nightmare, 67.
Northern Crown, 208, 211, 230.
Novelists, theme for, 193, 228.
Nucleus, 11, 19, 216. (See *Comets, Corona*.)

OCEANS, 179.
Old Astronomy, 199, 203, 233. (See *New*.)
Organisms in sun (*q. v.*), 13.
Orion, 238, 239, 247.
Oxygen, 73.

PACIFIC OCEAN, 180.
Palinurus, 243.
Parable, 224.
Paris: Observatory, 42, 233, 247; Exposition, 112.
Parker's Lens, 103.
Peirce, Professor, 44.
Pennsylvania Coal, 97.
Penumbra, 11, 19, 20.
Perpignan, France, 42.
Perseus, 196.
Persian Rugs, 70.
Philadelphia, 88.
Philosopher's Stone, 92.
Phœbus, 34.
Phosphorus, 221.
Photographic Plate, 71.
Photography, 9, 19, 128, 236, 237, 241, 244, 247, 248; rapid, 242.
Photometer, 56, 108.
Photometry, 230.
Photosphere, 7, 17, 64.
Pickering, Professor, 132, 227, 228, 246.
Pico Summit, 148.
Piffard, Dr. H. G., 245.
Pike's Peak, 50, 53-57, 60.
Pilgrim Fathers, 5.
Pine-boughs, 25.

Pine-trees, 60, 72.
Pittsburg Observations, 18, 19. (See *Allegheny, Langley.*)
Planetoids, 196, 197, 229.
Planets: condition, 97; pulverized, 100; and moon, 117-174; isolated, 176. (See *Jupiter, Mars, Mercury, Saturn, Sirius, Stars.*)
Plants, 72, 73. (See *Flowers.*)
Plato Crater, 147, 148, 151, 152.
Pleiades, 17, 231, 245. (See *Lost.*)
Plume, The, 19, 23, 24, 55.
Pointers, 208. (See *Dipper.*)
Poison, 222.
Polariscope, 49.
Polarization, 18.
Polarizing Eye-piece, 14, 18.
Polar Star, 230. (See *Great Bear.*)
Polyp, 152.
Pores, 24.
Pouillet's Invention, 93.
Printing, indebtedness to the sun, 74.
Prism, 63, 64. (See *Colors, Scarlet.*)
Proctor's Observations, 14, 59, 69, 87.
Prospero's Wand, 221.
Ptolemy, 155, 161.
Pyramids, 99, 117, 233, 234. (See *Egypt.*)
Pyrheliometer, 93.

RACE, simile, 179.
Radiant Energy, 71, 74; rate, 104.
Radiation, 101, 109.
Railway Explosion, 182, 183.
Railway, The, 156.
Rain, 111.
Rainbow, 70.
Ranyard's Photographs, 50.
Red Sea, 116.
Reflection, 216.
Repulsive Force, 215.
Ribbons, 70, 236.
Rifts, 163, 164.
Rings, 123, 124, 152, 155. (See *Saturn.*)
Rockets, 67, 68.
Rocky Mountains, 88, 89, 180.
Roman Boy, 34.
Rope, 20, 26.
Rose-leaf, 63, 70. (See *Leaves.*)
Rowland's Photographs, 237.
Ruskin, quoted, 29.
Russia, 134.

Rutherfurd Photographs, 8, 9, 137, 143, 155, 234.

SAL-AMMONIAC, 14, 25.
Salisbury Plain, 1, 2.
Sandstone, 192.
Saturn, 118, 119, 121, 123, 124, 127-129, 136, 215.
Saturnian Dwarfs, 223, 224.
Saul, comparison, 77.
Saxon Forefathers, 1, 2. (See *Britain.*)
Scarlet, 67. (See *Colors.*)
Schwabe, Hofrath, 76, 77, 87.
Scott, Sir Walter, quoted, 92.
Screen, 10, 35-37.
Seas, lunar (*q. v.*), 143.
Secchi, Father, 14, 15, 24, 25, 29, 30, 43, 59, 235.
Segmentatious, 30, 31.
Self-luminosity, 215.
Sextant, 224.
Shadows. (See *Lunar.*)
Shakspeare, quoted, 60, 220.
Sheaves, 68.
Shelbyville, 42, 43.
Sherman, observations at, 88.
Ship, comparison, 133. (See *Steamer.*)
Shooting-stars, 35, 193, 196, 198, 199. (See *Falling, Meteors.*)
Sicily, 50. (See *Etna.*)
Siemens, Sir William, 111.
Sierra Nevada, 151, 160, 180.
Signal Service, 90.
Silicon, 107.
Sirius, 179, 222-224, 236-239.
Slits, 59, 63, 64.
Smoked Glass, 11.
Snow-flakes, 19, 35.
Snow-like Forms, 25.
Sodium, 237.
Solar Engine, 75, 109.
Solar Light (*q. v.*), 13.
Solar Physics, 4, 12, 14. (See *Sun.*)
Solar System, 228, 229.
South America (*q. v.*), 80.
South Carolina, meteors, 194, 195. (See *Charleston.*)
Southern Cross, 234.
Space, 181, 211, 224, 227, 229.
Spain, expedition, 44.

INDEX. 259

Sparks, 107, 108.
Spectra, 231, 237.
Spectres, 54, 55. (See *Brocken.*)
Spectroscope, 7, 50, 59, 61, 63, 64, 130, 176, 198, 219, 222, 233-235, 240.
Spectrum, 65, 235.
Spectrum Analysis, 12.
Speculations, 193.
Spinning-wheel, 115.
Springfield Observations, 44.
Spurs, 208, 212, 215.
Star of Bethlehem, 229. (See *Tycho.*)
Stars: new study, 3; location, 4; size, 4, 230; seen in darkness, 35; self-shining suns, 35, 118; host, 117; variety, 118; five, 118; elements, atmosphere, 179; showers (see *Meteors*), 195; seen through comet, 212, 215; chapter, 221-250; analysis, children, 222; distances, 223; intervals, 224, 227, 229; colors (*q. v.*), glory, 227; new, fading, 230; double, 233; relation to man (*q. v.*), 233; fixed, 233; changing place, 234; mass, 237; ages, 238; photographed, 244, 247; chart, 247; death, 248. (See *Falling, Planets, Shooting.*)
Steam, 74, 75.
Steamers, 21, 73, 115.
Steel, melted, 104-108. (See *Iron.*)
Stellar Spectra (*q. v.*), 222, 236, 237, 244, 245.
Stevenson, George, 111.
Stewart's Observations, 88.
Stonehenge, 1-3.
Stones: from heaven, 175, 176, 186, 187, 191, 193; Iowa, 199, 200. (See *Meteorites.*)
Stonyhurst Records, 88.
Sumbawa Observations, 181.
Sunbeams: lifting power, 72; Laputa, 73; printing, 74; motes, 215. (See *Light.*)
Sun: practical observations in Washington, 2, 3; new study, 3; surroundings, 4, 35-69; distance, 4-6; size, 5, 6; a private, 6; views, 6-12, 15, 16, 20; details, 7; fire, 8, 91, 92; telescopic view, 8; axis, 9; revolutions, 10; surface, 17; paper record, 18; heat (*q. v.*) and eye, 19; drawings exaggerated, 29, 30; something brighter, 32; atmosphere, 33, 34; slits, 59; miniature, 64; flames (*q. v.*), 69; energy, 70-116 (see *Heat*); versatile aid, 74; children, 75, 222; shrinkage, 99; ground up, 100; emissive power, 104; constitution and appearance, 111; god, 116; self-shining, 118; studied from mountains, 167; affected by dust (*q. v.*), 185; and comet, 216; elements, 233; a star, 237; life, 238; candle, 249; anecdote, 250. (See *Solar.*)
Sunrise, 234.
Sunset, 181, 182. (See *Twilight.*)
Suns: millions, 224; dwindling, 227; periods, 241.
Sun-spots, 1-34 *passim;* ancient, 8; early observations, 8; changing, 9; great, 10, 20, 24; individuality, darker, 11; leaves (*q. v.*), 11, 12; how observed, 18, 19; typical, 21, 22; relative size, 20; hook-shaped (see *Plume*), 24; signs of chaos, 27; double, 32; weather, 76, 90; periodicity, 76-78; temperature, 83; records, 85; variations, 87; (1870), 9, 15, 16, 20; (1873), 20-24; (1875), 23, 28, 30; (1876), 30, 32; (1882), 80, 83-86, 90.
Superga, 38.
Swift, Dean, 73, 131, 132. (See *Gulliver.*)
Sword Meteor (*q. v.*), 175.

TACCHINI'S INVESTIGATIONS, 43, 49, 62, 66, 68.
Tail, 215, 216. (See *Comets.*)
Tau, 71.
Taylor, Bayard, 139.
Telephone, 84, 89.
Telescopes: many, 17; best, 134; alone, 227, 230; use, 233, 234.
Temperature, 101, 102, 108; of space, 224, 227.
Terminator, 147.
Thermometer, 71, 93, 102; low, 160, 163.
Time, small divisions, 241.
Tippoo Saib, 221.
Total Eclipse (*q. v.*), 39-48 *passim,* 55, 59.
Trees, lacking, 168.
Tribune, The New York, 84.
Trinity Church, 72.
Trocadéro, 112.
Trouvelot, E. L., 119, 123, 225.
Turin, 38.
Twilight, small, 38.
Tycho, 144, 229. (See *Star.*)
Tyndall, 98.

UMBRA, 11, 12, 19, 20.
United States, comparison, 24.
Uranus, 3, 196.

VAPOR, 28.
Vega, 235, 246.
Vegetables, 74.
Veils, 14, 17.
Venus, 118.
Vernier, 3.
Vesuvius: crater, 155, 157; eruption, 181, 183. (See *Naples*.)
Vibrations, 72.
Victoria, 115.
Viscous Fluid, 26.
Vital Force, 14.
Vogel, H. C., 64, 66.
Voids, 181, 227.
Volcanoes, 27, 28; in moon, 167, 193.

WANDERING STAR, 101. (See *Comets, Falling*.)
Washington: Observatory, 2, 86-88; telescope, 122; Monument, 182.
Water, 152; in man, 221.
Waterloo, 80.
Water-wheel, 111.
Watson's Observations, 49.

Wheat, prices, 79. (See *Breadstuffs, Corn, Grain, Sun-spots*.)
Wheel, comparison, 10.
Whirlpools, 28, 31.
Whirlwinds, 23, 31.
White Light (*q. v.*), 48, 62, 63.
Whitney, Mount, 177.
Willow-leaves (*q. v.*), 11, 12, 14.
Wing, simile, 215.
Winlock, Professor, 44.
Withered Surfaces, 168, 171.
Wood-engraving, 50.
Worlds and Clouds, 249.
Wrinkles, 172.

XERES, Spain (*q. v.*), 44, 53.

YOUNG, PROFESSOR: spectroscope, 44, 50, 65, 234; observations, 56, 59, 61, 68, 69; magnetism, 87, 88; radiation, 101.

ZODIACAL LIGHT, 55.

www.ingramcontent.com/pod-product-compliance
Lightning Source LLC
Chambersburg PA
CBHW032001230426
43672CB00010B/2232